Mometrix
TEST PREPARATION

Common Core Grade 6 Mathematics Secrets Study Guide

I0892280

DEAR FUTURE EXAM SUCCESS STORY

First of all, **THANK YOU** for purchasing Mometrix study materials!

Second, congratulations! You are one of the few determined test-takers who are committed to doing whatever it takes to excel on your exam. **You have come to the right place.** We developed these study materials with one goal in mind: to deliver you the information you need in a format that's concise and easy to use.

In addition to optimizing your guide for the content of the test, we've outlined our recommended steps for breaking down the preparation process into small, attainable goals so you can make sure you stay on track.

We've also analyzed the entire test-taking process, identifying the most common pitfalls and showing how you can overcome them and be ready for any curveball the test throws you.

Standardized testing is one of the biggest obstacles on your road to success, which only increases the importance of doing well in the high-pressure, high-stakes environment of test day. Your results on this test could have a significant impact on your future, and this guide provides the information and practical advice to help you achieve your full potential on test day.

Your success is our success

We would love to hear from you! If you would like to share the story of your exam success or if you have any questions or comments in regard to our products, please contact us at **800-673-8175** or **support@mometrix.com**.

Thanks again for your business and we wish you continued success!

Sincerely,
The Mometrix Test Preparation Team

Need more help? Check out our flashcards at:
http://MometrixFlashcards.com/CCSS

TABLE OF CONTENTS

Introduction

Thank you for purchasing this resource! You have made the choice to prepare yourself for a test that could have a huge impact on your future, and this guide is designed to help you be fully ready for test day. Obviously, it's important to have a solid understanding of the test material, but you also need to be prepared for the unique environment and stressors of the test, so that you can perform to the best of your abilities.

For this purpose, the first section that appears in this guide is the **Secret Keys**. We've devoted countless hours to meticulously researching what works and what doesn't, and we've boiled down our findings to the five most impactful steps you can take to improve your performance on the test. We start at the beginning with study planning and move through the preparation process, all the way to the testing strategies that will help you get the most out of what you know when you're finally sitting in front of the test.

We recommend that you start preparing for your test as far in advance as possible. However, if you've bought this guide as a last-minute study resource and only have a few days before your test, we recommend that you skip over the first two Secret Keys since they address a long-term study plan.

If you struggle with **test anxiety**, we strongly encourage you to check out our recommendations for how you can overcome it. Test anxiety is a formidable foe, but it can be beaten, and we want to make sure you have the tools you need to defeat it.

Secret Key #1 – Plan Big, Study Small

There's a lot riding on your performance. If you want to ace this test, you're going to need to keep your skills sharp and the material fresh in your mind. You need a plan that lets you review everything you need to know while still fitting in your schedule. We'll break this strategy down into three categories.

Information Organization

Start with the information you already have: the official test outline. From this, you can make a complete list of all the concepts you need to cover before the test. Organize these concepts into groups that can be studied together, and create a list of any related vocabulary you need to learn so you can brush up on any difficult terms. You'll want to keep this vocabulary list handy once you actually start studying since you may need to add to it along the way.

Time Management

Once you have your set of study concepts, decide how to spread them out over the time you have left before the test. Break your study plan into small, clear goals so you have a manageable task for each day and know exactly what you're doing. Then just focus on one small step at a time. When you manage your time this way, you don't need to spend hours at a time studying. Studying a small block of content for a short period each day helps you retain information better and avoid stressing over how much you have left to do. You can relax knowing that you have a plan to cover everything in time. In order for this strategy to be effective though, you have to start studying early and stick to your schedule. Avoid the exhaustion and futility that comes from last-minute cramming!

Study Environment

The environment you study in has a big impact on your learning. Studying in a coffee shop, while probably more enjoyable, is not likely to be as fruitful as studying in a quiet room. It's important to keep distractions to a minimum. You're only planning to study for a short block of time, so make the most of it. Don't pause to check your phone or get up to find a snack. It's also important to **avoid multitasking**. Research has consistently shown that multitasking will make your studying dramatically less effective. Your study area should also be comfortable and well-lit so you don't have the distraction of straining your eyes or sitting on an uncomfortable chair.

The time of day you study is also important. You want to be rested and alert. Don't wait until just before bedtime. Study when you'll be most likely to comprehend and remember. Even better, if you know what time of day your test will be, set that time aside for study. That way your brain will be used to working on that subject at that specific time and you'll have a better chance of recalling information.

Finally, it can be helpful to team up with others who are studying for the same test. Your actual studying should be done in as isolated an environment as possible, but the work of organizing the information and setting up the study plan can be divided up. In between study sessions, you can discuss with your teammates the concepts that you're all studying and quiz each other on the details. Just be sure that your teammates are as serious about the test as you are. If you find that your study time is being replaced with social time, you might need to find a new team.

Secret Key #2 – Make Your Studying Count

You're devoting a lot of time and effort to preparing for this test, so you want to be absolutely certain it will pay off. This means doing more than just reading the content and hoping you can remember it on test day. It's important to make every minute of study count. There are two main areas you can focus on to make your studying count.

Retention

It doesn't matter how much time you study if you can't remember the material. You need to make sure you are retaining the concepts. To check your retention of the information you're learning, try recalling it at later times with minimal prompting. Try carrying around flashcards and glance at one or two from time to time or ask a friend who's also studying for the test to quiz you.

To enhance your retention, look for ways to put the information into practice so that you can apply it rather than simply recalling it. If you're using the information in practical ways, it will be much easier to remember. Similarly, it helps to solidify a concept in your mind if you're not only reading it to yourself but also explaining it to someone else. Ask a friend to let you teach them about a concept you're a little shaky on (or speak aloud to an imaginary audience if necessary). As you try to summarize, define, give examples, and answer your friend's questions, you'll understand the concepts better and they will stay with you longer. Finally, step back for a big picture view and ask yourself how each piece of information fits with the whole subject. When you link the different concepts together and see them working together as a whole, it's easier to remember the individual components.

Finally, practice showing your work on any multi-step problems, even if you're just studying. Writing out each step you take to solve a problem will help solidify the process in your mind, and you'll be more likely to remember it during the test.

Modality

Modality simply refers to the means or method by which you study. Choosing a study modality that fits your own individual learning style is crucial. No two people learn best in exactly the same way, so it's important to know your strengths and use them to your advantage.

For example, if you learn best by visualization, focus on visualizing a concept in your mind and draw an image or a diagram. Try color-coding your notes, illustrating them, or creating symbols that will trigger your mind to recall a learned concept. If you learn best by hearing or discussing information, find a study partner who learns the same way or read aloud to yourself. Think about how to put the information in your own words. Imagine that you are giving a lecture on the topic and record yourself so you can listen to it later.

For any learning style, flashcards can be helpful. Organize the information so you can take advantage of spare moments to review. Underline key words or phrases. Use different colors for different categories. Mnemonic devices (such as creating a short list in which every item starts with the same letter) can also help with retention. Find what works best for you and use it to store the information in your mind most effectively and easily.

Secret Key #3 – Practice the Right Way

Your success on test day depends not only on how many hours you put into preparing, but also on whether you prepared the right way. It's good to check along the way to see if your studying is paying off. One of the most effective ways to do this is by taking practice tests to evaluate your progress. Practice tests are useful because they show exactly where you need to improve. Every time you take a practice test, pay special attention to these three groups of questions:

- The questions you got wrong
- The questions you had to guess on, even if you guessed right
- The questions you found difficult or slow to work through

This will show you exactly what your weak areas are, and where you need to devote more study time. Ask yourself why each of these questions gave you trouble. Was it because you didn't understand the material? Was it because you didn't remember the vocabulary? Do you need more repetitions on this type of question to build speed and confidence? Dig into those questions and figure out how you can strengthen your weak areas as you go back to review the material.

Additionally, many practice tests have a section explaining the answer choices. It can be tempting to read the explanation and think that you now have a good understanding of the concept. However, an explanation likely only covers part of the question's broader context. Even if the explanation makes perfect sense, **go back and investigate** every concept related to the question until you're positive you have a thorough understanding

As you go along, keep in mind that the practice test is just that: practice. Memorizing these questions and answers will not be very helpful on the actual test because it is unlikely to have any of the same exact questions. If you only know the right answers to the sample questions, you won't be prepared for the real thing. **Study the concepts** until you understand them fully, and then you'll be able to answer any question that shows up on the test.

It's important to wait on the practice tests until you're ready. If you take a test on your first day of study, you may be overwhelmed by the amount of material covered and how much you need to learn. Work up to it gradually.

On test day, you'll need to be prepared for answering questions, managing your time, and using the test-taking strategies you've learned. It's a lot to balance, like a mental marathon that will have a big impact on your future. Like training for a marathon, you'll need to start slowly and work your way up. When test day arrives, you'll be ready.

Start with the strategies you've read in the first two Secret Keys—plan your course and study in the way that works best for you. If you have time, consider using multiple study resources to get different approaches to the same concepts. It can be helpful to see difficult concepts from more than one angle. Then find a good source for practice tests. Many times, the test website will suggest potential study resources or provide sample tests.

Practice Test Strategy

If you're able to find at least three practice tests, we recommend this strategy:

Untimed and Open-Book Practice

Take the first test with no time constraints and with your notes and study guide handy. Take your time and focus on applying the strategies you've learned.

Timed and Open-Book Practice

Take the second practice test open-book as well, but set a timer and practice pacing yourself to finish in time.

Timed and Closed-Book Practice

Take any other practice tests as if it were test day. Set a timer and put away your study materials. Sit at a table or desk in a quiet room, imagine yourself at the testing center, and answer questions as quickly and accurately as possible.

Keep repeating timed and closed-book tests on a regular basis until you run out of practice tests or it's time for the actual test. Your mind will be ready for the schedule and stress of test day, and you'll be able to focus on recalling the material you've learned.

Secret Key #4 – Pace Yourself

Once you're fully prepared for the material on the test, your biggest challenge on test day will be managing your time. Just knowing that the clock is ticking can make you panic even if you have plenty of time left. Work on pacing yourself so you can build confidence against the time constraints of the exam. Pacing is a difficult skill to master, especially in a high-pressure environment, so **practice is vital**.

Set time expectations for your pace based on how much time is available. For example, if a section has 60 questions and the time limit is 30 minutes, you know you have to average 30 seconds or less per question in order to answer them all. Although 30 seconds is the hard limit, set 25 seconds per question as your goal, so you reserve extra time to spend on harder questions. When you budget extra time for the harder questions, you no longer have any reason to stress when those questions take longer to answer.

Don't let this time expectation distract you from working through the test at a calm, steady pace, but keep it in mind so you don't spend too much time on any one question. Recognize that taking extra time on one question you don't understand may keep you from answering two that you do understand later in the test. If your time limit for a question is up and you're still not sure of the answer, mark it and move on, and come back to it later if the time and the test format allow. If the testing format doesn't allow you to return to earlier questions, just make an educated guess; then put it out of your mind and move on.

On the easier questions, be careful not to rush. It may seem wise to hurry through them so you have more time for the challenging ones, but it's not worth missing one if you know the concept and just didn't take the time to read the question fully. Work efficiently but make sure you understand the question and have looked at all of the answer choices, since more than one may seem right at first.

Even if you're paying attention to the time, you may find yourself a little behind at some point. You should speed up to get back on track, but do so wisely. Don't panic; just take a few seconds less on each question until you're caught up. Don't guess without thinking, but do look through the answer choices and eliminate any you know are wrong. If you can get down to two choices, it is often worthwhile to guess from those. Once you've chosen an answer, move on and don't dwell on any that you skipped or had to hurry through. If a question was taking too long, chances are it was one of the harder ones, so you weren't as likely to get it right anyway.

On the other hand, if you find yourself getting ahead of schedule, it may be beneficial to slow down a little. The more quickly you work, the more likely you are to make a careless mistake that will affect your score. You've budgeted time for each question, so don't be afraid to spend that time. Practice an efficient but careful pace to get the most out of the time you have.

Secret Key #5 – Have a Plan for Guessing

When you're taking the test, you may find yourself stuck on a question. Some of the answer choices seem better than others, but you don't see the one answer choice that is obviously correct. What do you do?

The scenario described above is very common, yet most test takers have not effectively prepared for it. Developing and practicing a plan for guessing may be one of the single most effective uses of your time as you get ready for the exam.

In developing your plan for guessing, there are three questions to address:

- When should you start the guessing process?
- How should you narrow down the choices?
- Which answer should you choose?

When to Start the Guessing Process

Unless your plan for guessing is to select C every time (which, despite its merits, is not what we recommend), you need to leave yourself enough time to apply your answer elimination strategies. Since you have a limited amount of time for each question, that means that if you're going to give yourself the best shot at guessing correctly, you have to decide quickly whether or not you will guess.

Of course, the best-case scenario is that you don't have to guess at all, so first, see if you can answer the question based on your knowledge of the subject and basic reasoning skills. Focus on the key words in the question and try to jog your memory of related topics. Give yourself a chance to bring the knowledge to mind, but once you realize that you don't have (or you can't access) the knowledge you need to answer the question, it's time to start the guessing process.

It's almost always better to start the guessing process too early than too late. It only takes a few seconds to remember something and answer the question from knowledge. Carefully eliminating wrong answer choices takes longer. Plus, going through the process of eliminating answer choices can actually help jog your memory.

Summary: Start the guessing process as soon as you decide that you can't answer the question based on your knowledge.

How to Narrow Down the Choices

The next chapter in this book (**Test-Taking Strategies**) includes a wide range of strategies for how to approach questions and how to look for answer choices to eliminate. You will definitely want to read those carefully, practice them, and figure out which ones work best for you. Here though, we're going to address a mindset rather than a particular strategy.

Your odds of guessing an answer correctly depend on how many options you are choosing from.

Number of options left	5	4	3	2	1
Odds of guessing correctly	20%	25%	33%	50%	100%

You can see from this chart just how valuable it is to be able to eliminate incorrect answers and make an educated guess, but there are two things that many test takers do that cause them to miss out on the benefits of guessing:

- Accidentally eliminating the correct answer
- Selecting an answer based on an impression

We'll look at the first one here, and the second one in the next section.

To avoid accidentally eliminating the correct answer, we recommend a thought exercise called **the $5 challenge**. In this challenge, you only eliminate an answer choice from contention if you are willing to bet $5 on it being wrong. Why $5? Five dollars is a small but not insignificant amount of money. It's an amount you could afford to lose but wouldn't want to throw away. And while losing $5 once might not hurt too much, doing it twenty times will set you back $100. In the same way, each small decision you make—eliminating a choice here, guessing on a question there—won't by itself impact your score very much, but when you put them all together, they can make a big difference. By holding each answer choice elimination decision to a higher standard, you can reduce the risk of accidentally eliminating the correct answer.

The $5 challenge can also be applied in a positive sense: If you are willing to bet $5 that an answer choice *is* correct, go ahead and mark it as correct.

Summary: Only eliminate an answer choice if you are willing to bet $5 that it is wrong.

Which Answer to Choose

You're taking the test. You've run into a hard question and decided you'll have to guess. You've eliminated all the answer choices you're willing to bet $5 on. Now you have to pick an answer. Why do we even need to talk about this? Why can't you just pick whichever one you feel like when the time comes?

The answer to these questions is that if you don't come into the test with a plan, you'll rely on your impression to select an answer choice, and if you do that, you risk falling into a trap. The test writers know that everyone who takes their test will be guessing on some of the questions, so they intentionally write wrong answer choices to seem plausible. You still have to pick an answer though, and if the wrong answer choices are designed to look right, how can you ever be sure that you're not falling for their trap? The best solution we've found to this dilemma is to take the decision out of your hands entirely. Here is the process we recommend:

Once you've eliminated any choices that you are confident (willing to bet $5) are wrong, select the first remaining choice as your answer.

Whether you choose to select the first remaining choice, the second, or the last, the important thing is that you use some preselected standard. Using this approach guarantees that you will not be enticed into selecting an answer choice that looks right, because you are not basing your decision on how the answer choices look.

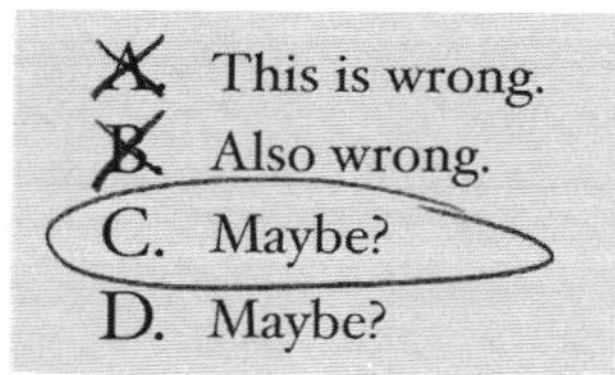

This is not meant to make you question your knowledge. Instead, it is to help you recognize the difference between your knowledge and your impressions. There's a huge difference between thinking an answer is right because of what you know, and thinking an answer is right because it looks or sounds like it should be right.

Summary: To ensure that your selection is appropriately random, make a predetermined selection from among all answer choices you have not eliminated.

Test-Taking Strategies

This section contains a list of test-taking strategies that you may find helpful as you work through the test. By taking what you know and applying logical thought, you can maximize your chances of answering any question correctly!

It is very important to realize that every question is different and every person is different: no single strategy will work on every question, and no single strategy will work for every person That's why we've included all of them here, so you can try them out and determine which ones work best for different types of questions and which ones work best for you.

Question Strategies

☑ Read Carefully

Read the question and the answer choices carefully. Don't miss the question because you misread the terms. You have plenty of time to read each question thoroughly and make sure you understand what is being asked. Yet a happy medium must be attained, so don't waste too much time. You must read carefully and efficiently.

☑ Contextual Clues

Look for contextual clues. If the question includes a word you are not familiar with, look at the immediate context for some indication of what the word might mean. Contextual clues can often give you all the information you need to decipher the meaning of an unfamiliar word. Even if you can't determine the meaning, you may be able to narrow down the possibilities enough to make a solid guess at the answer to the question.

☑ Prefixes

If you're having trouble with a word in the question or answer choices, try dissecting it. Take advantage of every clue that the word might include. Prefixes can be a huge help. Usually, they allow you to determine a basic meaning. *Pre-* means before, *post-* means after, *pro-* is positive, *de-* is negative. From prefixes, you can get an idea of the general meaning of the word and try to put it into context.

☑ Hedge Words

Watch out for critical hedge words, such as *likely, may, can, often, almost, mostly, usually, generally, rarely*, and *sometimes*. Question writers insert these hedge phrases to cover every possibility Often an answer choice will be wrong simply because it leaves no room for exception. Be on guard for answer choices that have definitive words such as *exactly* and *always*.

☑ Switchback Words

Stay alert for *switchbacks*. These are the words and phrases frequently used to alert you to shifts in thought. The most common switchback words are *but, although*, and *however*. Others include *nevertheless, on the other hand, even though, while, in spite of, despite*, and *regardless of*. Switchback words are important to catch because they can change the direction of the question or an answer choice.

✓ Face Value

When in doubt, use common sense. Accept the situation in the problem at face value. Don't read too much into it. These problems will not require you to make wild assumptions. If you have to go beyond creativity and warp time or space in order to have an answer choice fit the question, then you should move on and consider the other answer choices. These are normal problems rooted in reality. The applicable relationship or explanation may not be readily apparent, but it is there for you to figure out. Use your common sense to interpret anything that isn't clear.

Answer Choice Strategies

✓ Answer Selection

The most thorough way to pick an answer choice is to identify and eliminate wrong answers until only one is left, then confirm it is the correct answer. Sometimes an answer choice may immediately seem right, but be careful. The test writers will usually put more than one reasonable answer choice on each question, so take a second to read all of them and make sure that the other choices are not equally obvious. As long as you have time left, it is better to read every answer choice than to pick the first one that looks right without checking the others.

✓ Answer Choice Families

An answer choice family consists of two (in rare cases, three) answer choices that are very similar in construction and cannot all be true at the same time. If you see two answer choices that are direct opposites or parallels, one of them is usually the correct answer. For instance, if one answer choice says that quantity *x* increases and another either says that quantity *x* decreases (opposite) or says that quantity *y* increases (parallel), then those answer choices would fall into the same family. An answer choice that doesn't match the construction of the answer choice family is more likely to be incorrect. Most questions will not have answer choice families, but when they do appear, you should be prepared to recognize them.

✓ Eliminate Answers

Eliminate answer choices as soon as you realize they are wrong, but make sure you consider all possibilities. If you are eliminating answer choices and realize that the last one you are left with is also wrong, don't panic. Start over and consider each choice again. There may be something you missed the first time that you will realize on the second pass.

✓ Avoid Fact Traps

Don't be distracted by an answer choice that is factually true but doesn't answer the question. You are looking for the choice that answers the question. Stay focused on what the question is asking for so you don't accidentally pick an answer that is true but incorrect. Always go back to the question and make sure the answer choice you've selected actually answers the question and is not merely a true statement.

✓ Extreme Statements

In general, you should avoid answers that put forth extreme actions as standard practice or proclaim controversial ideas as established fact. An answer choice that states the "process should be used in certain situations, if…" is much more likely to be correct than one that states the "process should be discontinued completely." The first is a calm rational statement and doesn't even make a definitive, uncompromising stance, using a hedge word *if* to provide wiggle room, whereas the second choice is far more extreme.

✓ Benchmark

As you read through the answer choices and you come across one that seems to answer the question well, mentally select that answer choice. This is not your final answer, but it's the one that will help you evaluate the other answer choices. The one that you selected is your benchmark or standard for judging each of the other answer choices. Every other answer choice must be compared to your benchmark. That choice is correct until proven otherwise by another answer choice beating it. If you find a better answer, then that one becomes your new benchmark. Once you've decided that no other choice answers the question as well as your benchmark, you have your final answer.

✓ Predict the Answer

Before you even start looking at the answer choices, it is often best to try to predict the answer. When you come up with the answer on your own, it is easier to avoid distractions and traps because you will know exactly what to look for. The right answer choice is unlikely to be word-for-word what you came up with, but it should be a close match. Even if you are confident that you have the right answer, you should still take the time to read each option before moving on.

General Strategies

✓ Tough Questions

If you are stumped on a problem or it appears too hard or too difficult, don't waste time. Move on! Remember though, if you can quickly check for obviously incorrect answer choices, your chances of guessing correctly are greatly improved. Before you completely give up, at least try to knock out a couple of possible answers. Eliminate what you can and then guess at the remaining answer choices before moving on.

✓ Check Your Work

Since you will probably not know every term listed and the answer to every question, it is important that you get credit for the ones that you do know. Don't miss any questions through careless mistakes. If at all possible, try to take a second to look back over your answer selection and make sure you've selected the correct answer choice and haven't made a costly careless mistake (such as marking an answer choice that you didn't mean to mark). This quick double check should more than pay for itself in caught mistakes for the time it costs.

✓ Pace Yourself

It's easy to be overwhelmed when you're looking at a page full of questions; your mind is confused and full of random thoughts, and the clock is ticking down faster than you would like. Calm down and maintain the pace that you have set for yourself. Especially as you get down to the last few minutes of the test, don't let the small numbers on the clock make you panic. As long as you are on track by monitoring your pace, you are guaranteed to have time for each question.

✓ Don't Rush

It is very easy to make errors when you are in a hurry. Maintaining a fast pace in answering questions is pointless if it makes you miss questions that you would have gotten right otherwise. Test writers like to include distracting information and wrong answers that seem right. Taking a little extra time to avoid careless mistakes can make all the difference in your test score. Find a pace that allows you to be confident in the answers that you select.

⊘ Keep Moving

Panicking will not help you pass the test, so do your best to stay calm and keep moving. Taking deep breaths and going through the answer elimination steps you practiced can help to break through a stress barrier and keep your pace.

Final Notes

The combination of a solid foundation of content knowledge and the confidence that comes from practicing your plan for applying that knowledge is the key to maximizing your performance on test day. As your foundation of content knowledge is built up and strengthened, you'll find that the strategies included in this chapter become more and more effective in helping you quickly sift through the distractions and traps of the test to isolate the correct answer.

Now that you're preparing to move forward into the test content chapters of this book, be sure to keep your goal in mind. As you read, think about how you will be able to apply this information on the test. If you've already seen sample questions for the test and you have an idea of the question format and style, try to come up with questions of your own that you can answer based on what you're reading. This will give you valuable practice applying your knowledge in the same ways you can expect to on test day.

Good luck and good studying!

Mathematics

RATIO

A ratio is a direct comparison of two parts of a whole. This is not the same as comparing each part to the whole, as in the case of fractions. For example, for every 3 girls in Maggie's class there are 5 boys. The ratio of girls to boys is a direct comparison of the number of girls and boys, and is therefore 3:5. (Note how this is different from a fraction: 3/8 of the students in Maggie's class are girls, which relates the number of one part, girls, to the whole, students.) Any ratio can be written in three ways: with a colon, as a fraction, or with the word "to". For example, the ratio of girls to boys in Maggie's class can be written as either "3:5", "$\frac{3}{5}$", or "3 to 5". All three ways are read as "three to five".

The following shows how ratios can describe various situations:

- Every car in the parking lot has four wheels. - "Every car in the parking lot has four wheels," means the ratio of cars to wheels is 1:4, because for every one car there are four wheels.
- There are two eyes on every face. - "There are two eyes on every face," means the ratio of eyes to faces is 2:1, because for every two eyes there is one face.
- Every full deck of cards has four aces. - "Every full deck of cards has four aces," means either that the ratio of decks to aces is 1:4, because for each deck there are four aces, or that the ratio of cards to aces is 52:4, because for every set of 52 cards (1 deck), there are 4 aces.
- A recipe calls for two cups of flour for every two dozen cookies. - "A recipe calls for two cups of flour for every two dozen cookies," means the ratio of cups of flour to cookies is 2:24, because for every two cups of flour used 24 (two dozen) cookies will be made.

UNIT RATE

Unit rate expresses a quantity of one thing in terms of one unit of another. For example, if you travel 30 miles every two hours, a unit rate expresses this comparison in terms of one hour: in one hour you travel 15 miles, so your unit rate is 15 miles per hour. Other examples are how much one ounce of food costs (price per ounce), or figuring out how much one egg costs out of the dozen (price per 1 egg, instead of price per 12 eggs). The denominator of a unit rate is always 1. Unit rates are used to compare different situations to solve problems. For example, to make sure you get the best deal when deciding which kind of soda to buy, you can find the unit rate of each. If Soda #1 costs $1.50 for a 1-liter bottle, and soda #2 costs $2.75 for a 2-liter bottle, it would be a better deal to buy Soda #2, because its unit rate is only $1.375 per 1-liter, which is cheaper than Soda #1. Unit rates can also help determine the length of time a given event will take. For example, if you can paint 2 rooms in 4.5 hours, you can determine how long it will take you to paint 5 rooms by solving for the unit rate per room and then multiplying that by 5.

EXAMPLE PROBLEM 1

At the store you see two different bags of candy for sale. Bag A has 32 pieces of candy in it and costs \$2.10. Bag B has 50 pieces of candy in it and costs \$3.50. Find the unit cost per one piece of candy from each bag and determine which is the better deal.

One piece of candy in Bag A costs about \$0.065, or 6.5 cents. This can be found be taking the total price, \$2.10, and dividing it by the number of pieces of candy, 32, to determine the cost for one piece of candy: $\frac{2.10}{32} = 0.06562$, or about \$0.065. One piece of candy in Bag B costs \$0.07, or 7 cents. This can be found in the same way, dividing the total cost, \$3.50, by the number of pieces of candy, 50: $\frac{3.5}{50} = 0.07$, or \$0.07. Since a piece of candy in Bag B is slightly more expensive, it is the better deal to buy Bag A.

EXAMPLE PROBLEM 2

You decide to bake oatmeal cookies for a bake sale. Your recipe calls for 5 cups of flour, 2 cups of sugar, 1 cup of butter, and 3 cups of oats. However, you only have 1 cup of oats. You decide to make as many cookies as you can with the 1 cup of oats that you have. Find how much of each ingredient you will need to make that many cookies.

You will need to use $1\,{}^2/_3$ cups of flour, ${}^2/_3$ cup of sugar, and ${}^1/_3$ cup of butter. These can be found by finding the unit rate of each per 1 cup of oats. If there are 5 cups of flour needed for every 3 cups of oats, then $\frac{5}{3} = 1\,{}^2/_3$ cups of flour needed for every 1 cup of oats. Similarly, if there needs to be 2 cups of sugar for every 3 cups of oats, then there needs to be ${}^2/_3$ cup of sugar for every 1 cup of oats. And finally, if there is 1 cup of butter needed for every 3 cups of oats, then there will need to be ${}^1/_3$ cup of butter for every 1 cup of oats.

EXAMPLE PROBLEM 3

Janice made \$40 during the first 5 hours she spent babysitting. She will continue to earn money at this rate until she finishes babysitting in 3 more hours. Find how much money Janice earned babysitting and how much she earns per hour.

Janice will earn \$64 babysitting in her 8 total hours (adding the first 5 hours to the remaining 3 gives the 8 hour total). This can be found by setting up a proportion comparing money earned to babysitting hours. Since she earns \$40 for 5 hours and since the rate is constant, she will earn a proportional amount in 8 hours: $\frac{40}{5} = \frac{x}{8}$. Cross-multiplying will yield $5x = 320$, and division by 5 shows that $x = 64$.

Janice earns \$8 per hour. This can be found by taking her total amount earned, \$64, and dividing it by the total number of hours worked, 8. Since $\frac{64}{8} = 8$, Janice makes \$8 in one hour. This can also be found by finding the unit rate, money earned per hour: $\frac{64}{8} = \frac{x}{1}$. Since cross-multiplying yields $8x = 64$, and division by 8 shows that $x = 8$, Janice earns \$8 per hour.

EXAMPLE PROBLEM 4

The McDonalds are taking a family road trip, driving 300 miles to their cabin. It took them 2 hours to drive the first 120 miles. They will drive at the same speed all the way to their cabin. Find the speed at which the McDonalds are driving and how much longer it will take them to get to their cabin.

The McDonalds are driving 60 miles per hour. This can be found by setting up a proportion to find the unit rate, the number of miles they drive per one hour: $\frac{120}{2} = \frac{x}{1}$. Cross-multiplying yields $2x = 120$ and division by 2 shows that $x = 60$.

Since the McDonalds will drive this same speed for the remaining miles, it will take them another 3 hours to get to their cabin. This can be found by first finding how many miles the McDonalds have left to drive, which is 300 – 120 = 180. The McDonalds are driving at 60 miles per hour, so a proportion can be set up to determine how many hours it will take them to drive 180 miles: $\frac{180}{x} = \frac{60}{1}$. Cross-multiplying yields $60x = 180$, and division by 60 shows that $x = 3$. This can also be found by using the formula $D = r \times t$ (or $Distance = rate \times time$), where $180 = 60 \times t$, and division by 60 shows that $t = 3$.

EXAMPLE PROBLEM 5

It takes Andy 10 minutes to read 6 pages of his book. He has already read 150 pages in his book that is 210 pages long. Find how long it takes Andy to read 1 page and also find how long it will take him to finish his book if he continues to read at the same speed.

It takes Andy 1 minute and 40 seconds to read one page in his book. This can be found by finding the unit rate per one page, by dividing the total time it takes him to read 6 pages by 6. Since it takes him 10 minutes to read 6 pages, $\frac{10}{6} = 1\,^2/_3$ minutes, which is 1 minute and 40 seconds.

It will take Andy another 100 minutes, or 1 hour and 40 minutes to finish his book. This can be found by first figuring out how many pages Andy has left to read, which is 210-150 = 60. Since it is now known that it takes him $1\,^2/_3$ minutes to read each page, then that rate must be multiplied by however many pages he has left to read (60) to find the time he'll need: $60 \times 1\,^2/_3 = 100$, so it will take him 100 minutes, or 1 hour and 40 minutes, to read the rest of his book.

EXAMPLE PROBLEM 6

Find the unit rate in the situations given:

A. Pizza Place is offering a deal for $40 for 8 pizzas. What is the cost of each pizza?

B. The ratio of flight attendants to airplanes is 42:14. How many flight attendants are on each airplane?

C. Milo has a pledge of $100 for the 25 mile walk-a-thon. How much will Milo earn for each mile?

D. It takes the cleaning service 8 hours to clean 20 garages. How long does it take to clean 1 garage?

A. The unit rate is \$5 per pizza. This is found by $\frac{\$40}{8} = \$5$.

B. The unit ratio is 3:1 which means there are 3 flight attendants for each airplane. This is found by $\frac{42}{14} = 3$.

C. The unit rate is \$4 per mile. This is found by $\frac{\$100}{25} = \$4$.

D. The unit rate is 0.4 hours, or 24 minutes, to clean one garage. This is found by $\frac{8\ hours}{20} = 0.4$ hours. This can be converted to minutes multiplying by a conversion factor: $0.4\ hours \times \frac{60\ minutes}{1\ hour} = 24\ minutes.$

PERCENT

Percent is one way of expressing what portion something is out of a whole. You can think of it like dividing the whole into 100 equal parts, called percents. The whole, all 100 parts, is called 100% (read as "one hundred percent"). Half of the whole is half (50) of the parts, so we say it is 50%, and so on. (Percent compares one quantity to another; sometimes neither number is necessarily 'whole' or 'all' of something.) To find what percent one number is of another, divide the first by the second, and multiply the answer by 100. For example, we can use percent to compare a part to a whole by asking how much money someone spends on their housing payment each month out of his or her total amount of money that month. If someone earns \$2000 every month and spends \$800 of that on their housing payment, then that person spends $\frac{800}{2000} \times (100) = 0.4 \times 100 = 40\%$ of his or her income on housing payments. Another example showing how percent can simply relate two numbers is comparing your age to your mother's age. If you are 12 years old and your mother is 40 years old, then you are $\frac{12}{40} \times 100 = 0.3 \times 100 = 30\%$ of her age.

EXAMPLE PROBLEM 1

Johnny got 80% of the questions correct on his math test. Find how many questions Johnny answered correctly if the test had 75 questions on it. Also, find what Johnny's new score would be if he was able to earn an extra 10% of his original score by doing corrections on the questions he missed.

Johnny got 60 questions correct. To find this, set up the percent calculation, leaving the number Johnny got right as *x*: $(\frac{x}{75} \times 100 = 80\%$. Multiply both sides by 75 and then divide both sides by 100 to solve for $x = \frac{80 \times 75}{100} = 60$. We say we are finding 80% of 75, which we now know means simply multiplying 75 by 0.80, which is 60.

Johnny's new score would be $^{66}/_{75}$, which is a score of 88%. This can be found in two ways. One is finding what number of questions is 10% of Johnny's original number right, or 10% of 60, which is 6 questions. If Johnny gets those 6 questions added back to his score, he would then have a total of 66 questions correct out of 75, which is $\frac{66}{75}$, which is 0.88, or 88%. The other way is by finding 10% of Johnny's original score of 80% and adding that to his original score: 10% of 80% = $0.10 \times 80\% = 8\%$, and 80% + 8% = 88%.

EXAMPLE PROBLEM 2

A sweater at a local department store is on sale for \$33. Find the original price of the sweater if it is marked 40% off the original price.

The sweater was originally \$55. To find this, set up a proportion to determine what number 33 is 60% of. We use 60% because if the sweater is 40% *off* we *subtracted* 40% from the price, so we have 60% *left*. The proportion will compare the percent over 100 to 33 over the original price: $\frac{60}{100} = \frac{33}{x}$. Cross-multiplying yields $60x = 3300$, and division by 60 shows that $x = 55$.

Greater or Equal To

Examples

For each situation listed below, determine which quantity is greater or if they are equal:

3 yards and 20 feet

70 minutes and 1 hour

188 pennies and 2 dollars

2 feet and 24 inches

1 kilometer and 100 meters

20 feet is greater. Since there are 3 feet in each yard, there are 9 feet in 3 yards, and 20 feet is greater than 9 feet.

70 minutes is greater, since there are only 60 minutes in one hour.

2 dollars is greater. Since there are 100 pennies in each dollar, there are 200 pennies in 2 dollars, which is greater than 188 pennies.

These quantities are equal, since there are 12 inches in each foot, and therefore 24 inches in 2 feet.

1 kilometer is greater, because in each kilometer there are 1000 meters, which is greater than 100 meters.

CONVERSIONS

EXAMPLE PROBLEM 1

Tony is going to buy new carpet for two rooms in his house. One room is $12\ feet \times 12\ feet$, and the other is $10\ feet \times 18\ feet$. The carpet Tony wants to buy is on sale for \$10 per square yard. How much will Tony pay for the amount of carpet he needs for both rooms?

Tony will pay \$360 for his carpet. First figure out how much carpet Tony needs to buy in square yards and then multiply that by \$10 per square yard.

Since Tony's room dimensions are in feet, they must first be converted to yards. The first room is 12 feet × 12 feet, and since there are 3 feet in each yard, that room is 4 yards × 4 yards (because ${}^{12}/_{3} = 4$). Similarly, the room that is 10 feet × 18 feet measures $3\,{}^{1}/_{3}$ yards × 6 yards (because ${}^{10}/_{3} = 3\,{}^{1}/_{3}$, and ${}^{18}/_{3} = 6$). So, the total square yardage of Tony's rooms is $4 \times 4 = 16$ square yards, plus $3\,{}^{1}/_{3} \times 6 = 20$ square yards, or 16 + 20 = 36 square yards. Since the carpet costs \$10 per square yard, then the total cost is $\$10 \times 36 = \$360$.

EXAMPLE PROBLEM 2

Tyson is driving home from work. He is driving 45 miles per hour and is 10 miles away from home. Find how many minutes it will take Tyson to get the rest of the way home if he continues to drive at the same speed.

It will take Tyson $13\,{}^{1}/_{3}$ minutes to get home. Convert Tyson's speed to miles per minute, using a conversion factor of 60 minutes in 1 hour: $\frac{45\ miles}{1\ hour} \times \frac{1\ hour}{60\ minutes} = 0.75\ {}^{miles}/_{minute}$. (The 'hours' units cancel and the answer is then in 'miles per minute'.) Because $distance = rate \times time$, $\frac{distance}{rate} = time$. Therefore, if Tyson has 10 miles left to go, dividing that by 0.75 miles per minute finds how many minutes it will take Tyson to get home. $\frac{10}{0.75} = 13\,{}^{1}/_{3}$, so it will take Tyson $13\,{}^{1}/_{3}$ minutes to get home.

EXAMPLE PROBLEM 3

When Mrs. Smith turned 50 years old she wanted to figure out how many minutes she had been alive. She determined that she had been alive for 25,000,000 minutes. Is she correct?

No, Mrs. Smith is not correct. She has been alive for 26,280,000 minutes. This is found by converting 50 years into minutes, using conversion factors from years to days to hours to minutes: $50\ \text{years} \times \frac{365\ days}{1\ year} \times \frac{24\ hours}{1\ day} \times \frac{60\ minutes}{1\ hour} = 26{,}280{,}000\ \text{minutes}$. The first multiplication operation will give 18,250, which is how many days Mrs. Smith has been alive (the 'years' units cancel). Multiplying that by $\frac{24\ hours}{1\ day}$ will cancel the 'days' units and yield 438,000 hours. Finally, multiplying that by $\frac{60\ minutes}{1\ hour}$ will cancel the 'hours' units and leave an answer of 26,280,000 minutes.

Division

Example Problem

Solve the following division problem:

$$9\overline{)9027}$$

The answer to the problem of 9027 divided by 9 is 1003.

$$\begin{array}{r} 1003 \\ 9\overline{)9027} \\ \underline{9000} \\ 0027 \\ \underline{0027} \\ 0 \end{array}$$

Of course, 9 divides into 9 exactly one time, so a 1 is placed on the answer line, and the product of 1 and 9 is placed under the thousands place of the dividend. (The remaining places are zero). 9 does not go into 0, so a 0 is placed on the answer line above 0. Since 9 also does not go into 2, another 0 is placed on the answer line. 9 goes into 27 three times, so a 3 is placed on the answer line above the 7. There is no remainder, because 9 goes into 27 exactly 3 times with nothing left over.

Decimals

Example Problem 1

Add the following decimals:

$$\begin{array}{l} 13.48 \\ \underline{+2.5\quad} \end{array}$$

The answer is 15.98.

$$\begin{array}{l} 13.48 \\ \underline{+2.5\quad} \\ 15.98 \end{array}$$

It is important to line up the digits correctly which can be done by making sure that the decimal points are lined up vertically, which results in lining up digits that have the same place values. Start by putting a decimal point in the answer line directly below the decimal points in the problem, and then add the digits in the column farthest to the right. There is no digit in 2.5 that lines up with the 8 in 13.48, so 0 is added to 8, and the answer, 8, is brought down below the addition line to the answer. Next, add the 4 and 5, as they are in the same place value (the tenths place in this case); put their sum, 9, in the tenths place of the answer. Adding the digits in the ones place puts a 5 in the ones place of the answer. Finally, the 1 in the tens place in 13.48 isn't added to any digit in 2.5, so the sum of 1 and 0 is put in the tens place of the answer, which makes the final answer 15.98.

EXAMPLE PROBLEM 2

Subtract the following decimals:

$$\begin{array}{r} 37.4 \\ \underline{-\ 5.32} \end{array}$$

The answer is 32.08.

$$\begin{array}{r} 37.4 \\ \underline{-\ 5.32} \\ 32.08 \end{array}$$

It is important to line up the digits correctly which can be done by making sure that the decimal points are lined up vertically, which results in lining up digits that have the same place values. Start by putting a decimal point in the answer line directly below the decimal points in the problem, and then subtract the digits in the column farthest to the right, the top minus the bottom. Since there is currently no digit above the 2, it is considered a 0, but since 2 cannot be subtracted from 0, borrow from the 4, making the 0 a 10 and making the 4 a 3. Since 2 can now be subtracted from 10, put the answer, 8, in the hundredths place of the answer. Now subtract 3 (remember: it's not a 4 anymore!) from 3, which is 0. In the ones place subtract 5 from 7, which is 2. And finally nothing is being subtracted from the 3 in the tens place in 37.4, so the 3 is brought down to the answer, making the final answer 32.08.

PRIME FACTORIZATION

Prime Factorization is breaking a number down into a list of its prime factors. For example, the prime factorization of 60 is $2 \times 2 \times 3 \times 5$, because 2, 3, and 5 are all prime numbers and $2 \times 2 \times 3 \times 5 = 60$. This can be found by first breaking down 60 into any two factors, perhaps 30 and 2. Since 2 is a prime number, it's already a prime factor, but 30 needs to be broken down into factors. 30 can be broken into 2 and 15, 3 and 10, or 6 and 5, but it does not matter which we use. If 3 and 10 are used, the factorization of 60 is $2 \times 3 \times 10$. Since 2 and 3 are prime but 10 is not, we have to break 10 down into 2 and 5, which now makes the factorization of $60 = 2 \times 3 \times 2 \times 5$. This is the prime factorization of 60 because these numbers are all prime. To report the answer in proper form put the numbers in order from least to greatest: the final answer for the prime factorization of 60 is $2 \times 2 \times 3 \times 5$.

GREATEST COMMON FACTOR

The greatest common factor is the greatest number that all numbers in a set are evenly divisible by. For example, the Greatest Common Factor of 36 and 48 is 12. One way to find this is by simply listing all the factors of 36 and 48; the common factors of 36 and 48 are 1, 2, 3, 4, 6, and 12, and 12 is the greatest of these. This can also be found using the prime factorization of both 36 and 48. The prime factorization of 36 is $2 \times 2 \times 3 \times 3$, and the prime factorization of 48 is $2 \times 2 \times 2 \times 2 \times 3$. Find how many of each prime factor these factorizations have in common: both factorizations have two 2's and one 3, so the Greatest Common Factor of 36 and 48 must be $2 \times 2 \times 3$, which equals 12.

EXAMPLES

The following sums can be expressed by factoring out a common factor:

15 + 21

16 + 20

50 + 80

27 + 63

6 + 52

18 + 66

'15 + 21' = 3(5 + 7)

'16 + 20' = 4(4 + 5)

'50 + 80' = 10(5 + 8)

'27 + 63' = 9(3 + 7)

'6 + 52' = 2(3 + 26)

'18 + 66' = 6(3 + 11)

These answers cannot be factored any further because there are no common factors left between the numbers inside the parentheses. In other words, the number that was factored out of each pair of numbers is that pair's greatest common factor.

LEAST COMMON MULTIPLE

The least common multiple is the smallest number that is evenly divisible by all numbers in a given set. For example, the least common multiple of 6 and 8 is 24. One way to find this is to list the common multiples of 6 and 8, 24, 48, 72, ...etc, and the least of these is 24. This can also be found using the prime factorization of both 6 and 8. The prime factorization of 6 is 2×3, and the prime factorization of 8 is $2 \times 2 \times 2$. For each factor, count the number of times it occurs in each factorization. Find the greatest number of times each factor shows up, and multiply each factor that many times. Since 2 shows up three times in 8's factorization, and 3 shows up once in 6's prime factorization, multiply $2 \times 2 \times 2 \times 3$, which equals 24.

HIGHEST AND LOWEST TEMPERATURE

EXAMPLE PROBLEM

At 1am in Anchorage, Alaska, it was -17° outside. The temperature fell a total of 10 degrees by sunrise, and then rose steadily a total of 35 degrees until 1pm, when the highest temperature of the day occurred. Find the lowest temperature during the night and the warmest temperature of the day.

The lowest temperature during the night was -27° and the warmest temperature of the day was 8°. Since the temperature falls 10° from -17°, subtract 10 from -17: -17 - 10 = -27, and the lowest temperature was -27°. From this lowest temperature it rises a total 35° which brings the thermometer up to 8° because -27 + 35 = 8. This can be thought of in two ways: first, because of the commutative property, -27 + 35 = 35 + -27, which is the same as 35 – 27, which equals 8. The other way is to recognize that the temperature must rise 27 degrees from -27° to get to 0°, but we know it rises another 8°, because the difference between 35 and 27 is 8.

VARIABLES, NUMBERS, AND OPERATION SYMBOLS

EXAMPLES

Use variables, numbers, and operation symbols to express the following:

3 less than a number ***n***

The quotient of 4 and a number ***x***

The difference between 10 and ***y***

The product of 12 and ***m***

The sum of ***x*** *and 4*

6 more than a number ***r***

"3 less than a number n" means taking 3 away from the value of n, so this is expressed as $n - 3$.

"The quotient of 4 and a number x" means dividing 4 by x, so we write $\frac{4}{x}$.

"The difference between 10 and y" means subtracting y from 10, so we write $10 - y$.

"The product of 12 and m" means multiplying 12 and m, so we write $12 \times m$, or simply $12m$.

"The sum of x and 4" means adding x and 4, so we write x +_4.

"6 more than a number r" means adding 6 to r, so we write $r + 6$.

Addition and Subtraction with Positives and Negatives

Example Problem 1

On Monday of last week, Molly's credit card had a balance of $50 owing on it. Throughout the week Molly made new charges on her credit card when she spent $75 at the grocery store on Tuesday, $30 on a haircut on Wednesday, and $15 at the movies on Friday. On Saturday she made a payment of $200 to her credit card and then on Sunday made another charge of $40 to buy new shoes. Determine whether Molly's credit card has a balance on Sunday night, and if so, how much it is.

Molly's credit card has a balance of $10 on Sunday night. This can be found by subtracting Molly's charges and adding her payment to her balance. A charge on a credit card, or balance, is a negative number because it is owed to the credit card company. A payment on the card gives the company money they are owed, so it is a positive number added to the balance to make that balance *less* negative. In Molly's case, she starts the week with a $50 balance, which is like starting with -50. The three charges on Tuesday, Wednesday, and Friday need to be subtracted from this, to get $-50 - 75 - 30 - 15 = -170$. When Molly makes a payment of $200, then, she only owed $170 on her credit card, so that -170 + 200 = positive 30. (Positive balances are money the company owes to Molly!) After that she makes another charge of $40, which is subtracted from 30 to get a final amount of -10. This means that at the end of the week Molly owes her credit card company $10, and we say her credit card has a balance of $10.

Example Problem 2

A helicopter is flying directly above a submarine. The submarine travels to 600 feet below sea level, while the helicopter reaches its flying height of 1050 feet. Find how far away the submarine and helicopter are from each other.

The helicopter and submarine are 1650 feet apart. This is found by adding their distances to sea level, because we can think of sea level as a value of zero in this situation. The helicopter is 1050 feet above sea level, which is like +1050, and its distance to sea level, 0, is 1050 feet. The submarine is 600 feet below sea level, which is like -600, and the distance from the submarine to sea level is 600 feet. The total distance between the helicopter and submarine, then, is 1050 + 600 = 1650 feet.

Properties

Example Problem 1

State which property is being used in each number sentence:

$$2x + y = y + 2x$$

$$5 \times (x + 1) = (5 \times x) + (5 \times 1)$$

$$3 \times 1 = 3$$

$$6 \times m \times n = m \times n \times 6$$

$$4 \times (5a) = (4a) \times 5$$

$$s + 0 = s$$

$$10 + (6 + 1) = (10 + 6) + 1$$

The Commutative Property of Addition is shown here, which states that you can add terms in any order.

The Distributive Property is shown here, which states that a number multiplied to an expression in parentheses must be multiplied to every term in the parentheses.

The Identity Property of Multiplication is shown here, which states that multiplying a number or term by 1 does not change its value.

The Commutative Property of Multiplication is shown here, which states that you can multiply terms in any order.

The Associative Property of Multiplication is shown here, which states that any group of numbers and/or variables can be grouped together in parentheses to be multiplied first before multiplying by the remaining numbers and/or variables.

"$s + 0 = s$" – The Identity Property of Addition is shown here, which states that adding 0 to any number or term does not change the value of that number or term.

The Associative Property of Addition is shown here, which states that any group of numbers and/or variables can be grouped together in parentheses to be added first before adding the remaining numbers and/or variables.

EXAMPLE PROBLEM 2

Simplify the following expressions and state what property, or properties, you used:

A. $3(2x - 1)$

B. $4x + y + 2x - 3y$

C. $2(5 + m) - 2m$

D. $7(2x)$

A. This simplifies to "$6x - 3$". The multiplication by 3 has to be distributed to both terms in the parentheses, which is the Distributive Property.

B. This simplifies to "$6x - 2y$", since combining like terms by adding them first ($4x + 2x = 6x$) and then adding those sums uses the Commutative Property of Addition.

C. This simplifies to 10. This is done using the Distributive Property to first multiply the 2 to each term in the parentheses to get "$10 + 2m - 2m$" and then combining like terms, using the Commutative Property of Addition. Since the like terms in this case are a positive $2m$ and a negative $2m$, they add to zero and therefore the answer is simply 10.

D. This simplifies to "$14x$" using the Associative Property of Multiplication.

Example Problem 3

Use the distributive property to match one expression in Column I with an equivalent expression in Column II.

Column I	Column II
A. 4(3x-1)	U. 10x+25y
B. 8(9+2x)	V. 12x-4
C. 36-2x	W. 3(4x-8y)
D. 4(4a+5)	X. 72+16x
E. 12x-24y	Y. 2(18-x)
F. 5(2x+5y)	Z. 16a+20

A is equivalent to V, by distributing the '4' into the parentheses.

B is equivalent to X by distributing the '8' into the parentheses.

C is equivalent to Y by distributing the '2' in Y back into the parentheses, or by factoring 2 from both the '36' and the '2x'.

D is equivalent to Z by distributing the '4' into the parentheses.

E is equivalent to W by distributing the '3' in W back into the parentheses, or by factoring '3' from both the '12x' and '24y'.

F is equivalent to U by distributing the '5' into the parentheses.

Proportions

Example Problem 1

Using the table below, find the values of a and b.

Gallons of Gas	Price in Dollars
2	a
5	20
b	32
9	36

The value of a is 8 and the value of b is also 8. They can both be found by setting up proportions. To find the value of a, a proportion can be set up comparing 2 and a with the values in the second row, 5 and 20: $\frac{2}{a} = \frac{5}{20}$. Cross multiplying yields $40 = 5a$, and division by 5 shows that $a = 8$.

To find the value of b, another proportion can be set up, comparing b and 32 with any other pair of data in the table. Using 9 and 36 yields the proportion $\frac{b}{32} = \frac{9}{36}$, and cross multiplying yields $36b = 288$. Division by 36 shows that $b = 8$.

Example Problem 2

Using the tables comparing Janaya's and Tim's running speeds, decide who will win a race that is 100 meters long.

Janaya	Tim

Distance	Time
10m	1.4 seconds
45m	6.3 seconds
85m	11.9 seconds

Distance	Time
12m	1.8 seconds
50m	7.5 seconds
85m	11.25 seconds

Janaya will win a 100m race against Tim, because Janaya will run 100m in 14 seconds, and Tim will run 100m in 15 seconds. Set up proportions to solve for what each runner's time will be for a 100m distance. To solve for Janaya's time, we can compare 100m with the unknown time, x, to any other pair of data in her table. If we use 10m and 1.4 seconds, the proportion $\frac{10}{1.4} = \frac{100}{x}$ yields $10x = 140$ after cross-multiplication and division by 10 yields $x = 14$. To solve for Tim's time, we can compare 100m with his unknown time, y, to any other pair of data in his table. If we use 50m and 7.5 seconds, the proportion $\frac{50}{7.5} = \frac{100}{y}$ will give $50y = 750$ and division by 50 will give $y = 15$. Since it will take Tim 1 second longer, Janaya will win.

COORDINATE PLANE

Four quadrants and the range of x and y-values

The four quadrants in a coordinate plane are called Quadrant I, Quadrant II, Quadrant III, and Quadrant IV. Quadrant I is the upper right quadrant of the plane and contains ordered pairs where $x > 0$ and $y > 0$. Quadrant II is the upper left quadrant of the plane and contains ordered pairs where $x < 0$ and $y > 0$. Quadrant III is the lower left quadrant of the plane and contains ordered pairs where $x < 0$ and $y < 0$. Quadrant IV is the lower right quadrant of the plane and contains ordered pairs where $x > 0$ and $y < 0$. The coordinate plane below shows the four quadrants:

EXAMPLE PROBLEM 1

Plot the data in the table below on a coordinate plane and find the value of y when $x = 1$.

X	Y
-4	-1
-2	1
0	3
2	5

Pairs of values (x, y) in the table are plotted on the coordinate plane below:

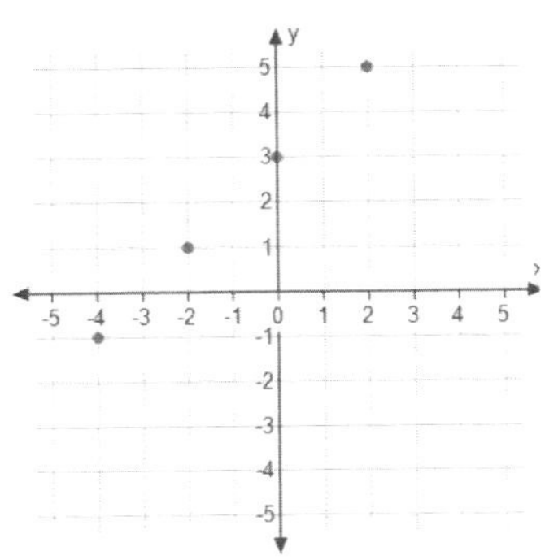

The value of y is 4 when $x = 1$. This can be found by looking at the graph or the table to determine what the pattern of the data is and where the point would lie when $x = 1$. The pattern in the graph and table is that y increases at a rate of 1 vertically for every 1 horizontally; in other words, for every time x increases by 1, y increases by 1 also. It can then be determined that since the y-value was 3 when the x-value was 0, the y-value will be 4 when the x-value is 1.

EXAMPLE PROBLEM 2

Plot the following points in a coordinate plane and describe each point's location:

Point A: (5, 0)

Point B: (-2, 2)

Point C: (-3, -1)

Point D: (3, 1)

Point E: (0, 0)

Point F: (0, -4)

Point G: (2, -3)

The graph below shows the location of the points:

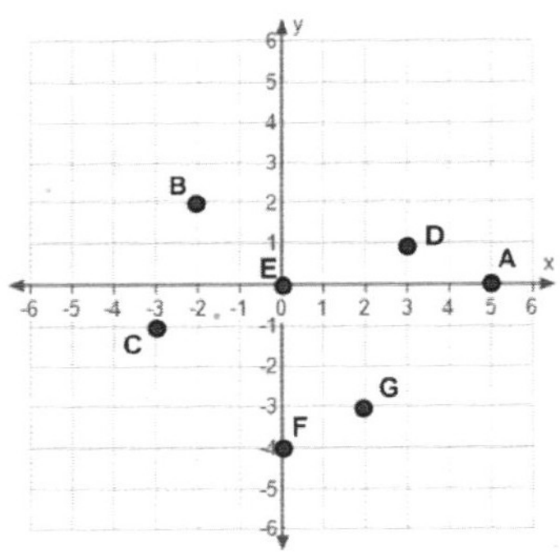

Point A is located on the $x - axis$, 5 units to the right of the origin. Point B is located in Quadrant II, 2 units to the left of the origin and 2 units up. Point C is located in Quadrant III, 3 units to the left of the origin and 1 unit down. Point D is located in Quadrant I, 3 units to the right of the origin and 1 unit up. Point E is located at the origin. Point F is located on the $y - axis$, 4 units down from the origin. Point G is located in Quadrant IV, 2 units to the right of the origin and 3 units down.

EXAMPLE PROBLEM 3

Find the perimeter of the rectangle in the coordinate plane below using the values of the coordinates of the vertices:

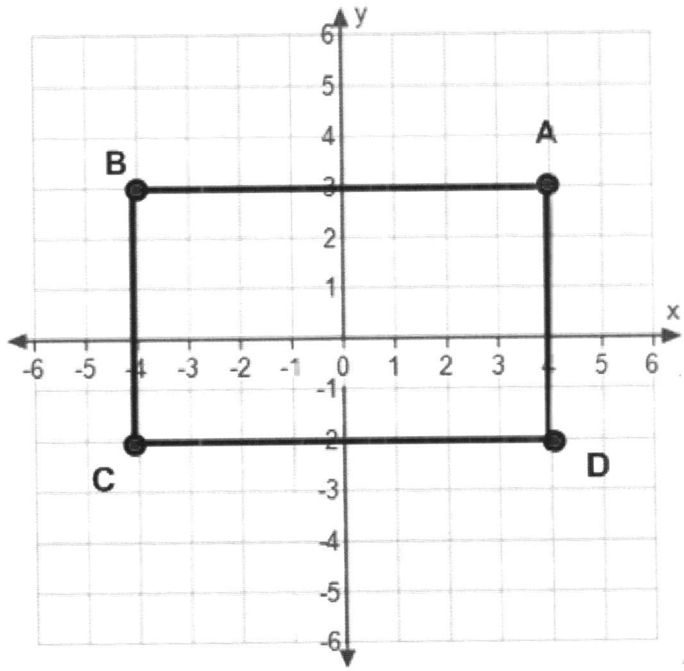

The perimeter of the rectangle is 26 units. This can be found by first finding the ordered pairs of each vertex of the rectangle, which are A= (4, 3), B= (-4, 3), C= (-4, -2), and D= (4, -2), and then using these to find the lengths of the four sides of the rectangle. Since each consecutive pair of vertices share either the same $x -$ coordinate or $y -$ coordinate, it is only necessary to consider the coordinates that are different in value. The distances from A to B and from C to D are each 8, which can be found by finding the distance between their $x -$ coordinates. The distance between 4 and -4 is 8, because 4 is 4 units away from zero and -4 is also 4 units away from zero which gives a total length of $4 + 4 = 8$. The distance from B to C and from D to A are each 5, which can be found by finding the distance between their $y -$ coordinates. The distance between 3 and -2 is 5, because 3 is 3 units away from zero and -2 is 2 units away from zero which gives a total length of $3 + 2 = 5$. Therefore the total perimeter of the rectangle is $8 + 8 + 5 + 5 = 26$ units.

EXAMPLE PROBLEM 4

Find the remaining ordered pairs of the vertices of a square in the coordinate plane below if one side length of the square is 5 units and it has a vertex in each quadrant.

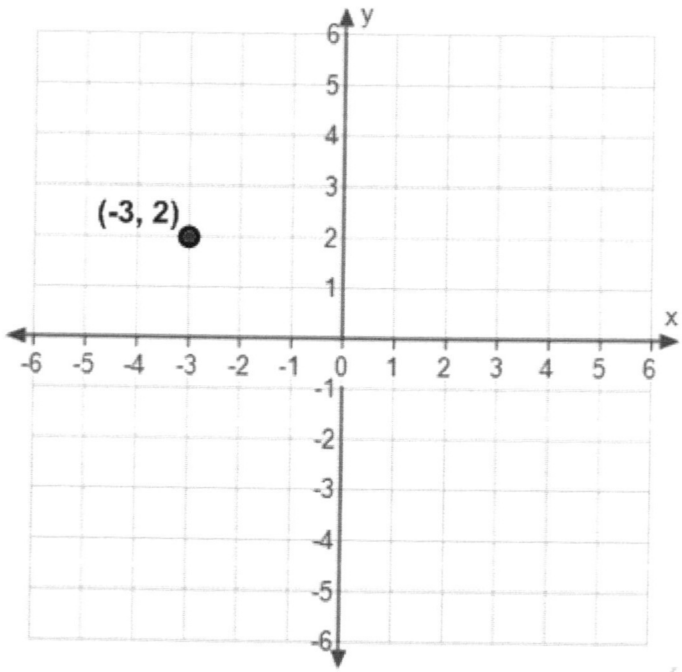

The remaining vertices have the ordered pairs of (-3, -3), (2, -3), and (2, 2). This is found by first figuring out what point in Quadrant III is 5 units away from the given point. It will have the same x −coordinate, so the value of y needs to be 5 units away from 2. Going 2 units away from 2 gets to 0, and then 3 more units must be traveled, getting to the point (-3, -3). Then figure out what point in Quadrant IV is 5 units away from (-3, -3). It will have the same y −coordinate, so the value of x needs to be 5 units away from -3. Going 3 units to the right of -3 gets to 0, and then 2 more units must be traveled, getting to the point (2, -3). Finally figure out what point in Quadrant I is 5 units away from (2, -3). It will have the same x −coordinate as the point in Quadrant IV, and the same y-coordinate as the point in Quadrant II, so the point must be (2, 2). This can be verified: the value of y needs to be 5 units away from -3. Going 3 units away from -3 gets to 0, and then 2 more units must be traveled, getting to the point (2, 2).

COORDINATE GRID

EXAMPLE PROBLEM 1

Your neighborhood can be represented by a coordinate grid, with each unit representing 1 block. Your house is located at (1,5), the park is located at (5,5), the drug store is located at (5,0), and the gas station is located at (1,0). Each morning you go for a run from your house along a route to the park, then the drug store, the gas station, and then back home. Plot this route on a coordinate grid and determine how many blocks you run.

You run 18 blocks on this route:

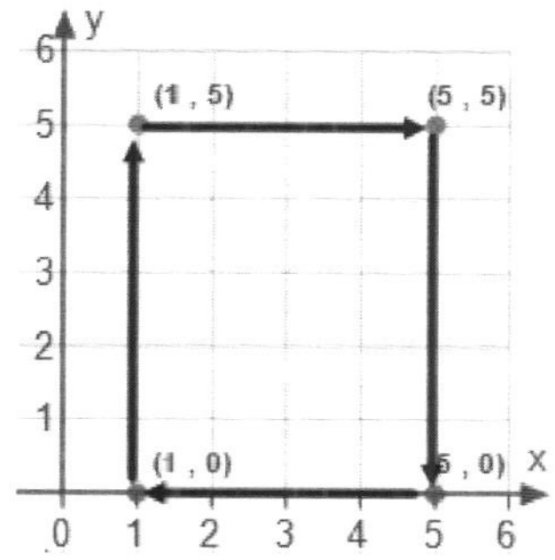

With each dot representing a location in your neighborhood, the route forms a rectangle. The first part of the route is 4 blocks to get from your house to the park, then 5 blocks to get from the park to the drug store, 4 more blocks to get from the drug store to the gas station, and finally 5 blocks to get from the gas station back to your house. Adding all these together gives you $4 + 5 + 4 + 5 = 18$ blocks. These distances can be found by either counting the number of squares on the grid between each location, or by finding the distance between the x or y coordinates (depending whether you're moving horizontally or vertically on the graph, respectively) of two points. For example, the distance between your house at (1,5) and the park at (5,5) is $5 - 1 = 4$ blocks, because you need to go 4 blocks over to get from the x -coordinate of 1 to the x-coordinate of 5, while not moving y-coordinates at all.

EXAMPLE PROBLEM 2

Platt Middle School was designed using a coordinate grid with each unit of measure representing 1 block. In the school, the library is located at (1, 4), the cafeteria is located at (-3, 4) and the gym is located at (1, -2). Plot these points on a coordinate grid and find the shortest distance a student needs to walk to get from gym class to the library and then to lunch.

The student will need to walk 10 blocks. This can be found by first finding the distance from the gym at (1, -2) to the library at (1, 4), and then the library to the cafeteria at (-3, 4). To get from the gym to the library the student will need to walk along a straight path along the x-coordinate of 1, from y-coordinates -2 to 4. The distance -2 is from 0 is 2, and the distance 4 is from 0 is 4. Therefore, the student must walk $2 + 4 = 6$ blocks from the gym to the library. Then the student will walk in a straight path along the y-coordinate of 4 from x-coordinates 1 to -3. The distance 1 is from 0 is 1, and the distance -3 is from 0 is 3. Therefore, the student must walk $1 + 3 = 4$ blocks from the library to lunch in the cafeteria. Thus, the student must walk a total of $6 + 4 = 10$ blocks to get from the gym to the library and then to lunch. The graph below shows the location of the places in the school and the student's path:

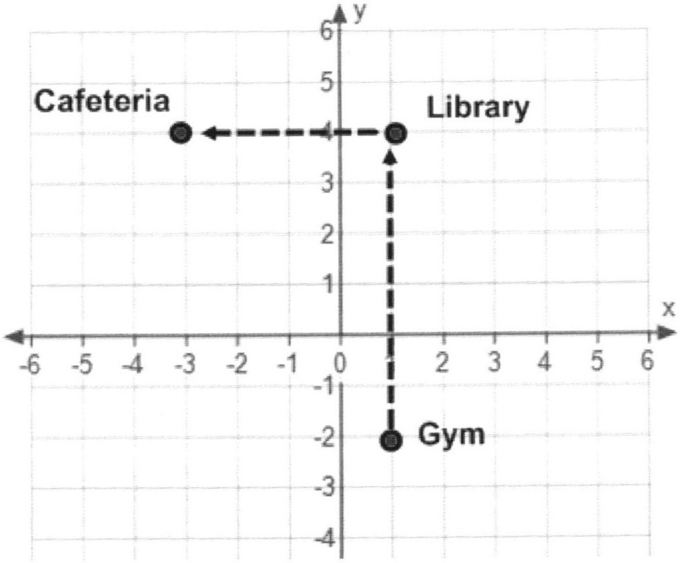

Example Problem 3

The downtown area of Cannonville has streets that follow a coordinate grid. The mayor is planning a parade through downtown and the city council has limited the length of the parade to 16 blocks. The mayor wants the parade to travel from the Zoo at (-5, -4) to the Park at (-3, -4), to the Bakery at (-3, 1), to the Library at (-1, 1), to the Post Office at (-1, -1), to City Hall at (2, -1), and ending at the Town Square at (2, 2). Plot this proposed route of the parade on a coordinate plane and determine if it will fit the requirement of the city council.

The distance of the proposed parade is 17 blocks so it will not make the 16 block requirement from the city council. This distance can be found by determining the distance between each pair of consecutive parade locations. Since each pair of consecutive locations share either an x or y −coordinate, the distance is found by calculating the distance between the different coordinates. When finding the distance between coordinates of opposite signs, first find the absolute value of each, which gives the distance of each from 0, and add them together. The distances are: **1.** Zoo to Park is 2 blocks, which is the distance between x −coordinates -5 and -3. **2.** Park to Bakery is 5 blocks, which is the distance between y-coordinates -4 and 1 because $|-4| + |1| = 5$. **3.** Bakery to the Library is 2 blocks, which is the distance between x −coordinates -3 and -1. **4.** Library to Post Office is 2 blocks, which is the distance between y-coordinates 1 and -1 because $|-1| + |1| = 2$. **5.** Post Office to City Hall is 3 blocks, which is the distance between$x$-coordinates -1 and 2 because $|-1| + |2| = 3$. **6.** City Hall to Town Square is 3 blocks, which is the distance between y-coordinates -1 and 2 because $|-1| + |2| = 3$. **7.** The total distance is then $2 + 5 + 2 + 2 + 3 + 3 = 17$.

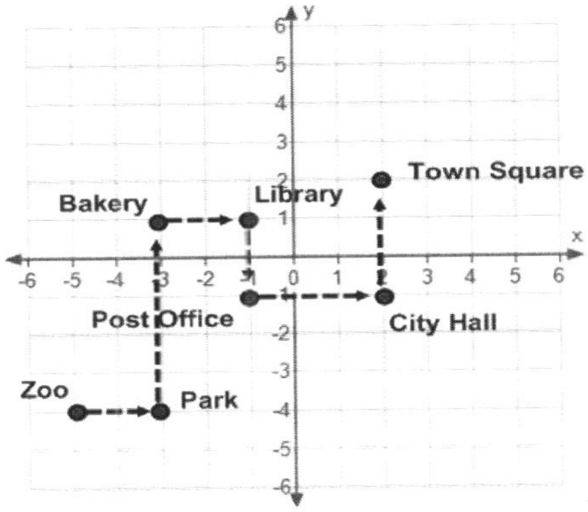

REFLECTIONS

EXAMPLE PROBLEM 1

Label the following points and their indicated reflections in a coordinate plane and give the coordinates of the reflections:

Point A: (4, 3) reflected across the $x-axis$

Point B: (1, 2) reflected across the $y-axis$

Point C: (-2, -5) reflected across the $y-axis$

Point D: (3, 1) reflected across the $x-axis$ and then the $y-axis$

The graph below shows the points and their reflections:

The reflection of Point A is labeled as A' and has coordinates (4, -3). Because it is reflected across the $x-axis$, the $y-value$ changes in sign.

The reflection of Point B is labeled as B' and has coordinates (-1, 2). Because it is reflected across the $y-axis$, the $x-value$ changes in sign.

The reflection of Point C is labeled as C' and has coordinates (2, -5). Because it is reflected across the $y-axis$, the $x-value$ changes in sign.

The reflection of Point D is labeled as D' and has coordinates (-3, -1). Because it is reflected across both the x and $y-axes$, both the x and $y-values$ change in sign.

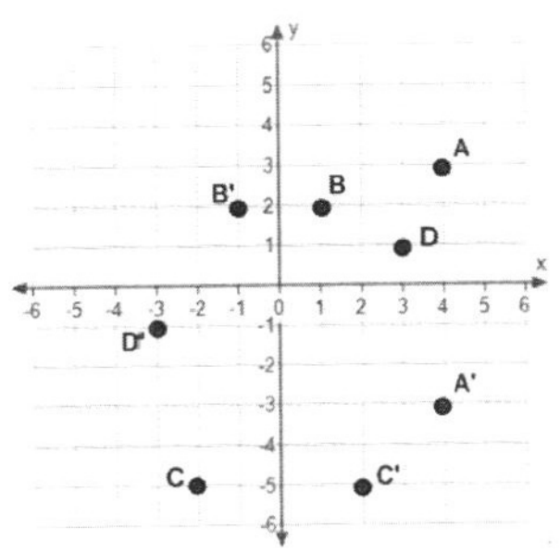

Example Problem 2

Identify the reflections that produce the second of each pair from the first of each pair:

(2, 6) and (-2, 6)

(-4, 0) and (4, 0)

(-1, -1) and (-1, 1)

(5, -7) and (-5, 7)

The reflection from (2, 6) to (-2, 6) is a reflection across the $y - axis$. This is because the $x - value$ has changed in sign.

The reflection from (-4, 0) to (4, 0) is a reflection across the $y - axis$. This is because the $x - value$ has changed in sign.

The reflection from (-1, -1) to (-1, 1) is a reflection across the $x - axis$. This is because the $y - value$ has changed in sign.

The reflection from (5, -7) to (-5, 7) is a reflection across both the x and $y - axis$. This is because both the x and $y - values$ have changed in sign.

Fractions

Example Problem 1

You and your best friend plan to split the remaining amount of a large pizza. If there is $\frac{3}{4}$ of the pizza left, find what fraction of the original pizza each of you will get and represent the solution with a diagram.

You each get $\frac{3}{8}$ of the original pizza. This is found by dividing $\frac{3}{4}$ by 2. When dividing with fractions, dividing by a number is the same as multiplying by the reciprocal of that number, so that $\frac{3}{4} \div 2 = \frac{3}{4} \times \frac{1}{2}$. Multiply the numerators together to obtain the numerator of the product, and multiply the denominators to get the denominator of the product: $\frac{3}{4} \times \frac{1}{2} = \frac{3}{8}$. This division problem can be represented by the diagram shown.

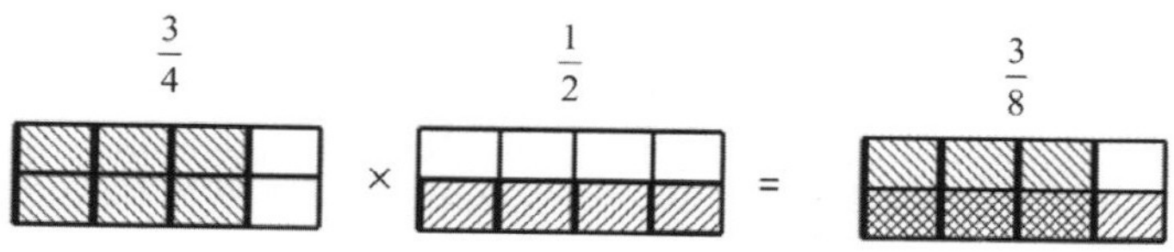

In the first part of the diagram, $\frac{3}{4}$ of the squares are shaded. In the second diagram, $\frac{1}{2}$ of the squares are shaded. Any squares that are shaded in both diagrams will be the answer. The overlap of the two diagrams shows that 3 of the 8 squares overlap, or $\frac{3}{8}$ of the whole. Or, consider dividing the first diagram, which has 6 out of the 8 squares shaded, by two. This would mean taking away half of the shaded squares, which would also result in 3 of the 8 squares shaded, or an answer of $\frac{3}{8}$.

Example Problem 2

Ms. Redding is going to frame two new rectangular posters she bought while on vacation. Before placing the order for the frames she needs to know the dimensions of each poster. When she bought the posters she was told that the area of each is $^2/_3$ square yards, the width of one is $^3/_4$ yards and the width of the other is $^1/_2$ yard. Find the length of each poster.

The lengths of the two posters are $^8/_9$ yards and $1\,^1/_3$ yards, respectively. Because $Area = length \times width$ for any rectangle, then $l = \frac{A}{w}$. To solve for the length of each rectangle, divide the area ($^2/_3$ square yards) by each width. When dividing by a fraction you must multiply by the reciprocal of that fraction.

For the first poster, $l = \frac{2}{3} \div \frac{3}{4} = \frac{2}{3} \times \frac{4}{3} = \frac{8}{9}$, and the length of the first poster is $^8/_9$ yards. For the second poster, $l = \frac{2}{3} \div \frac{1}{2} = \frac{2}{3} \times \frac{2}{1} = \frac{4}{3} = 1\,^1/_3$. and the length of the second poster is $1\,^1/_3$ yards.

Expressions

Example Problem 1

Simplify the following expressions:

$$(x+2) \times (x+2) \times (x+2)$$

$$(3-y) \times (3-y) \times (4x) \times (4x) \times (4x)$$

$$(5-m) \times (5-m)^2 \times (6-m)$$

$(x+2) \times (x+2) \times (x+2) = (x+2)^3$ This is because there are three instances of $(x+2)$ being multiplied together, which is the same as raising the quantity $(x+2)$ to the third power.

$(3-y) \times (3-y) \times (4x) \times (4x) \times (4x) = (3-y)^2 \times (4x)^3$ This is because there are two instances of $(3-y)$ being multiplied together, which is the same as raising the entity $(3-y)$ to the second power. That is then being multiplied by three instances of $(4x)$, which is the same as multiplying by $(4x)$ raised to the third power.

$(5-m) \times (5-m)^2 \times (6-m) = (5-m)^3 \times (6-m)$ This is because there are three instances of $(5-m)$ being multiplied together, because $(5-m)^2 = (5-m) \times (5-m)$.

EXAMPLE PROBLEM 2

Evaluate the following:

$$(3-1)+4^2\times 2$$

$$3^3+(3\times 3)$$

$$5^2-(6-2)^2$$

Evaluate these expressions following the order of operations: parentheses, exponents, multiplication/division, and addition/subtraction.

$(3-1)+4^2\times 2$ is equivalent to 34: $(3-1)+4^2\times 2=(2)+4^2\times 2=(2)+16\times 2=(2)+32=34$

$3^3+(3\times 3)$ is equivalent to 36: $3^3+(3\times 3)=3^3+9=27+9=36$

$5^2-(6-2)^2$ is equivalent to 9: $5^2-(6-2)^2=5^2-(4)^2=25-16=9$

EXAMPLE PROBLEM 3

Simplify the following expressions and identify the terms, products, factors, sums, quotients, and/or coefficients in each:

A. $4x+3x-2y$

B. $10(3+1)$

C. $\frac{10+5+1}{2}$

There are three terms in the original expression, $4x, 3x,$ and $-2y$. There are also three coefficients in the expression, 4, 3, and -2, respectively. Since $7x$ is the sum of $4x$ and $3x$, the expression simplifies to $7x-2y$.

There are two terms in the original expression, 3 and 1, and one coefficient, 10, multiplying the sum of the two terms. The sum of 3 and 1 is 4, and the product of 10 and 4 is 40. Therefore, 10 and 4 are factors of 40.

Since we can write this as $(10+5+1)\div 2$, there are four terms in the original expression, 10, 5, 1 and 2. The sum of 10, 5, and 1 is 16, and the quotient of 16 and 2 is 8. Therefore, 2 and 8 are factors of 16.

EXAMPLE PROBLEM 4

Write expressions to represent the following situations:

Turner has saved \$75 toward the purchase of an iPod. How much more money does he need to save?

This week Marcus mowed twice the amount of lawns he mowed last week. How many lawns did Marcus mow this week?

Jenny earns \$4 for every room she cleans in her house. How much money will she earn cleaning her whole house?

Turner has saved $75 toward the purchase of an iPod. The amount of money he still needs to save can be expressed by $C - 75$, representing the difference between total cost, C, of the iPod and what he already has.

This week Marcus mowed twice the amount of lawns he mowed last week. The number of lawns he mowed this week can be expressed as $2l$, representing two times the number of lawns mowed last week, l.

Jenny earns $4 for every room she cleans in her house. The amount of money she will earn cleaning her whole house can be expressed by $4r$, representing 4 times the number of rooms, r, in her whole house.

Example Problem 5

Write expressions to represent the following situations and then solve them:

Caroline has 3 more hours of homework tonight than last night. How much homework does Caroline have tonight if she had one hour of homework last night?

This month Fred will only work one-third of the hours he worked last month. How many hours will Fred work this month if he worked 150 hours last month?

The first part of this situation can be expressed by $h + 3$, representing the number of hours of homework Caroline has tonight in terms of 3 more than she had last night, h. If she had one hour of homework last night, then she has 4 hours of homework tonight, because $1 + 3 = 4$.

This can be expressed by $\frac{w}{3}$, representing the number of hours Fred will work this month in terms of one third of the number he worked last month, w. If he worked 150 hours last month, then he will work 50 hours this month, because $\frac{150}{3} = 50$.

Writing Equations

Example Problem 1

Write and solve equations that represent the following situations:

You are 16 years old and twice as old as your sister. How old is your sister?

Your mother is 5 times as old as your sister. How old is your mother?

Your cousin's age is the sum of your age and your sister's age. How old is your cousin?

This situation is represented by the equation $2x = 16$, where x represents your sister's age. Solve this equation by dividing both sides by 2, resulting in $x = 8$. Therefore, your sister is 8 years old.

Before we knew your sister's age, this was represented by $m = 5x$, where m is your mother's age and x is your sister's. Since it is now known that your sister is 8 years old, this situation is represented by the equation $5 \times 8 = m$. Solving this equation gives an answer of $5 \times 8 = 40$, so your mother is 40 years old.

Until we knew your sister's age, this was represented by $c = 16 + x$. Since your sister is 8 years, this situation can be represented by the equation $16 + 8 = c$, where c

represents your cousin's age. Solving this equation gives an answer of $16 + 8 = 24$, so your cousin is 24 years old.

EXAMPLE PROBLEM 2

Determine the equation that represents the cost, C, for any number of shirts, x, using the table of information below, and find how much it will cost to buy 12 shirts:

Number of Shirts	Cost (in dollars)
1	5
2	10
3	15
4	20

The equation that represents the situation is $C = 5x$. This is found by determining the relationship between the independent variable, which is the number of t-shirts, and the dependent variable, which is the cost. In the table it can be seen that for each shirt bought, the cost goes up by 5 dollars. It is then also seen that multiplying the number of shirts by 5 gives the cost, because $1 \times 5 = 5, 2 \times 5 = 10, 3 \times 5 = 15$, and $4 \times 5 = 20$. Therefore, for any x number of shirts, that number can just be multiplied by 5 to get the cost, which is represented by $C = 5x$.

It will cost \$60 to buy 12 shirts. This is found by using the equation that represents the data, substituting12 for x, the number of shirts: $C = 5 \times 12 = 60$.

EXAMPLE PROBLEM 3

Write the equation to represent the following table of values and graph them. Using the graph, predict the cost of mailing a 6 pound package.

The cost of mailing a package at the post office.	
Weight (in lbs)	Cost (in dollars)
1	1.5
2	3
3	4.5
4	6

The equation that represents the cost, C, in terms of the weight, w, of a package is $C = 1.5w$, because for each additional pound, the cost increases by $1.50 per pound. These values are graphed in the graph below:

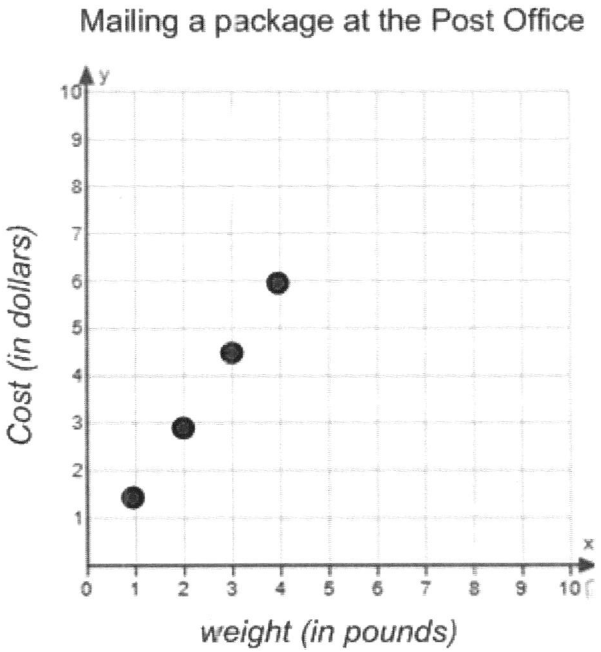

Looking at the graph, the points are going up at a constant rate of $1.50 for each pound. The next point would be at (5, 7.5), and then the next point would be at (6, 9). This means that the cost of mailing a 6 pound package would be $9.

Checking Equations

Example Problem 1

Match each expression in Column I with its equivalent expression in Column II. Verify that the expressions are equivalent by substituting any non-zero value for x.

Column I	Column II
A. $4x$	X. $x + 2x + x$
B. $10x + 12$	Y. $3x - x - x$
C. x	Z. $3x + 5 + 7x + 7$

A is equivalent to X. If $x = 3$, we see that $4x = 4 \times 3 = 12$, and $x + 2x + x = 3 + 2 \times 3 + 3 = 3 + 6 + 3 = 12$. Since the expressions have the same value when $x = 3$, the expressions are equivalent.

B is equivalent to Z. If $x = 2$, we see that $B = 10x + 12 = 10 \times 2 + 12 = 20 + 12 = 32$, and $Z = 3x + 5 + 7x + 7 = 3 \times 2 + 5 + 7 \times 2 + 7 = 6 + 5 + 14 + 7 = 32$. Since the expressions have the same value when $x = 2$, the expressions are equivalent.

C is equivalent to Y. If $x = 5$, we see that $x = 5$, and $3x - x - x = 3 \times 5 - 5 - 5 = 15 - 5 - 5 = 10 - 5 = 5$. Since the expressions have the same value when $x = 5$, the expressions are equivalent.

Example Problem 2

The equation $y = 2x + 5$ represents the line that contains all points (x, y) that satisfy the equation. Decide which, if any, of the following points lie on the line.

$$(1,8) \qquad (0,5) \qquad (2,9) \qquad (2,5)$$

The points (0,5) *and* (2,9) lie on the line. Any points that lie on the line have x and y values that make the equation true. Plug in the first value of each pair for x and the second value of each pair for y, and see whether the equation is true. For the point (0,5): $5 = 2 \times 0 + 5 = 0 + 5 = 5$, and the equation is still true. For the point (2,9): $9 = 2 \times 2 + 5 = 4 + 5 = 9$, and the equation is still true. These points make the equation true and lie on the line. The other two points do not work in the equation. For (1,8): $8 = 2 \times 1 + 5 = 2 + 5 \neq 7$. For (2,5): $5 = 2 \times 2 + 5 = 4 + 5 \neq 9$. (This could also be discovered by referring back to the point (2,9) that is on the line. Since y equals 9 when x is 2, the point (2,5) could be eliminated without testing it in the equation.)

Example Problem 3

Solve the inequality for a and check the answer by plugging in a possible value.

$$2a + 6 > 30$$

The solution is $a > 12$. This means that a can be any value greater than 12. This is solved the same way an equation is solved, doing whatever is done to one side of the inequality to the other:

$2a + 6 > 30$ (first subtract 6 from both sides)

$2a > 24$ (divide both sides by 2)

$$a > 12$$

To check the solution, any value that is greater than 12 can be chosen for a. If 20 is chosen, check to see that the left side of the inequality will have a greater value than 30:

$$\begin{aligned} 2 \times 20 + 6 &> 30 \\ 40 + 6 &> 30 \\ 46 &> 30 \end{aligned}$$

Because it is true that 46 is greater than 30, the inequality was solved correctly.

Example Problem 4

Find the solution to the following equation and check the solution by plugging it back into the equation:

$$2p - 11 = 19$$

The solution is $p = 15$. This is found by manipulating the equation to get p by itself on one side of the equal sign. This can be done best by getting rid of the numbers in the reverse order of the normal order of operations. Surrounding the p there is a 2 being multiplied to it and an 11 being subtracted from it. In order of operations, multiplication would be done first and subtraction after, so to undo this expression the subtraction must be handled first, then the multiplication:

$2p - 11 = 19$ (first add 11 to both sides to undo the subtraction by 11)

$2p = 30$ (then divide both sides by 2 to undo the multiplication by 2)

$p = 15$

By plugging the solution back into the original equation it can be verified: $2 \times 15 - 11 = 30 - 11 = 19 = 19$. Since both sides of the equation have a value of 19, the solution is correct.

TABLE REPRESENTING AN EQUATION

EXAMPLE

The equation $T = 1l$, represents T, the total time in minutes a racecar driver drives for l number of laps. Determine what the 1 in the equation means, and make a table of values that represents this equation and graph them.

The 1 in the equation represents the time the driver takes to complete one lap. This means that every lap takes the driver 1 minute. The table and graph below show this data:

Number of laps	Time (in minutes)
1	1
2	2
3	3
4	4

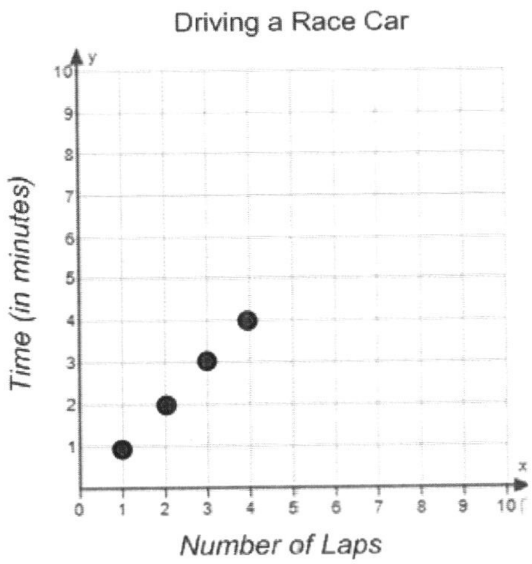

SURFACE AREA AND VOLUME

EXAMPLE PROBLEM 1

Find the Surface Area and Volume of a cube with side of length 3cm given the following formulas:

$SA = 6s^2$ *and* $V = s^3$

The Surface Area of the cube is 54 square cm and the Volume of the cube is 27 cubic cm. In a cube, the length, width, and height are all the same measurement, which is why the formulas only have one variable, s, which refers to the length of each side.

The answers are found by substituting 3 for s in each formula. Surface Area: $SA = 6 \times 3^2 = 6 \times 9 = 54cm^2$. Volume: $V = 3^3 = 27cm^3$.

EXAMPLE PROBLEM 2

Find the Surface Area and Volume of a cylinder with a height of 4cm and radius of 6cm, given the formulas below.

$SA = 2\pi r^2 + 2\pi rh$ and $V = \pi r^2 h$

The Surface Area of the cylinder is 120π square cm and the Volume of the cylinder is 144π cubic cm. Substitute 4 for h in each formula, since h represents the height, and 6 for r in each formula, since r represents the radius. Surface Area:

$$SA = 2\pi \times 6^2 + 2\pi \times 6 \times 4 = 2\pi \times 36 + 2\pi \times 24 = 72\pi + 48\pi = 120\pi\ cm^2.$$

Volume: $V = \pi \times 6^2 \times 4 = \pi \times 36 \times 4 = 144\pi\ cm^3$.

AREA

EXAMPLE PROBLEM 1

Find the area of the following isosceles trapezoid:

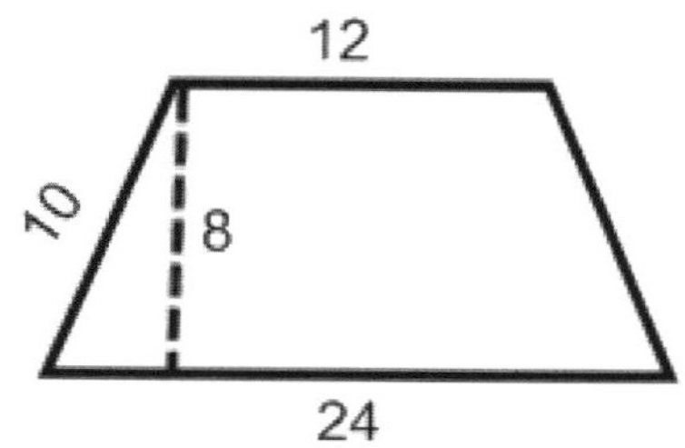

The area of the trapezoid is 144 square units and can be found by breaking the shape into two triangles and one rectangle, like so:

- Since the top base of the trapezoid has a length of 12 and is also the top of the rectangle, then the bottom of the rectangle also has a length of 12. Because the whole base had a length of 24, this leaves 12 for the two bases of the triangles. Dividing this remaining portion in half (because we are told the trapezoid is isosceles) gives each triangle a base of 6. The area of the rectangle is $12 \times 8 = 96$, and the area of each triangle is $\frac{6\times8}{2} = \frac{48}{2} = 24$. The total area is then $96 + 24 + 24 = 144$ square units.

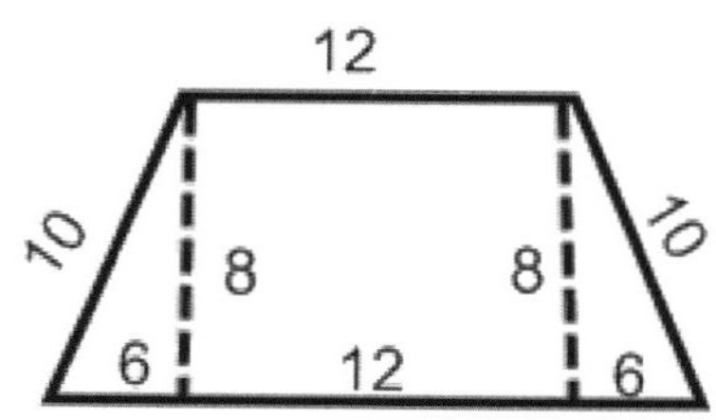

Example Problem 2

Mr. Byron is going to buy a new plot of land to build a cabin. The two plots of land are each triangles, but one is a right triangle and the other is an isosceles triangle. Decide which plot will give Mr. Byron the most land to build his cabin using the dimensions given below in feet:

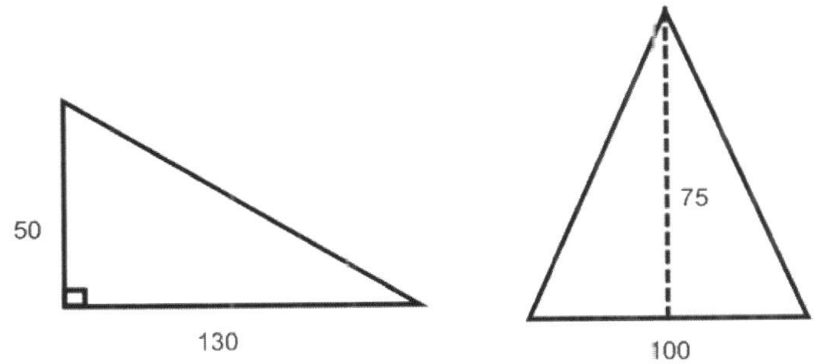

The isosceles triangle will give Mr. Byron the most land, because it has an area of 3750 square feet, while the right triangle only has an area of 3250 square feet. This is found by using the formula for the area of a triangle, $A = \frac{b \times h}{2}$. For the isosceles triangle, $A = \frac{100 \times 75}{2} = \frac{7500}{2} = 3750$ square feet, and for the right triangle, $A = \frac{130 \times 50}{2} = \frac{6500}{2} = 3250$ square feet. Therefore, the plot of land that is an isosceles triangle has an area that is 500 square feet larger than the right triangle, so Mr. Byron will want to purchase that plot.

Example Problem 3

Jeremy is staining the cement floor in his basement. He wants to stain the shape of a hexagon in the middle of the floor. He outlines a regular hexagon that has a side of length 6 feet on all sides and a height of 10.4 feet. Find the area Jeremy needs enough stain to cover.

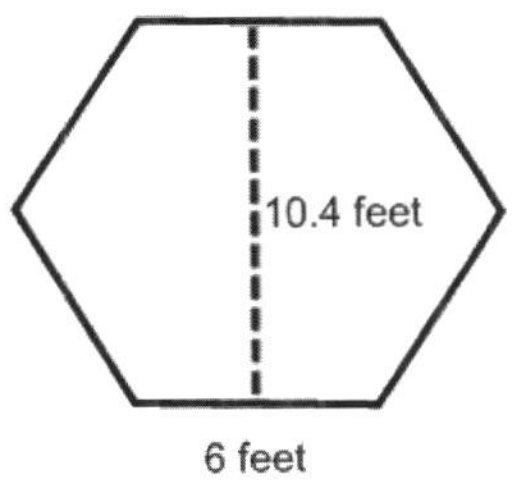

Jeremy will need enough stain to cover 93.6 square feet. This area of the hexagon is found by breaking the hexagon into 6 triangles. Because it is a regular hexagon, each side length is 6 and the height of one triangle is half the height of the whole hexagon or $\frac{10.4}{2} = 5.2$. Therefore all the triangles are congruent as seen in the diagram below:

Using the formula for area of a triangle, $A = \frac{b \times h}{2}$, each triangle has an area of $\frac{6 \times 5.2}{2} = \frac{31.2}{2} = 15.6$ square feet. Since each of the 6 triangles has the same area, the total area of the hexagon is then $15.6 \times 6 = 93.6$ square feet.

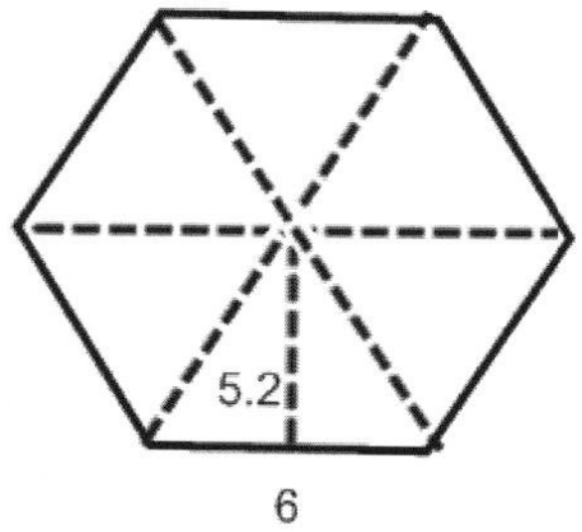

EXAMPLE PROBLEM 4

Find the area of the right triangle that has its longest side connecting points A (3, 8) and B (9, 4), and has one horizontal leg and one vertical leg.

The area of this triangle is 12 square units.

To solve for the area, it may be helpful to draw a diagram. Since we are told one leg is horizontal and the other vertical, the legs of the right triangle meet at the ordered pair (3, 4) (or at (9, 8) which would give the same area). The height of the triangle is the distance from (3, 4) to point A at (3, 8) which is 4 units. The base of the triangle is the distance from (3, 4) to point B at (9, 4) which is 6 units. The area of a triangle is found by the formula $A = \frac{b \times h}{2}$ and for this triangle the area is $A = \frac{6 \times 4}{2} = \frac{24}{2} = 12 \text{ units}^2$.

VOLUME

EXAMPLE PROBLEM 1

Ryland Raisin Company is creating new mini-boxes of raisins to sell at school lunches. They have decided to package them in a box that is a cube with dimensions of $\frac{4}{3}$ in $\times \frac{4}{3}$ in $\times \frac{4}{3}$ in. Find how many mini-boxes with side lengths of $\frac{1}{3}$ in can fit into a raisin box package and what volume the larger box can hold.

There are 64 cubes with side lengths of $\frac{1}{3}$ in that can fit into this new raisin box. The volume of the box is $\frac{64}{27} \text{in}^3 = 2\frac{10}{27} \text{in}^3$.

The number of $\frac{1}{3}$ in sided cubes that fit into the box can be found by creating a diagram of the box with its dimensions broken down into $\frac{1}{3}\,in$ portions as seen below:

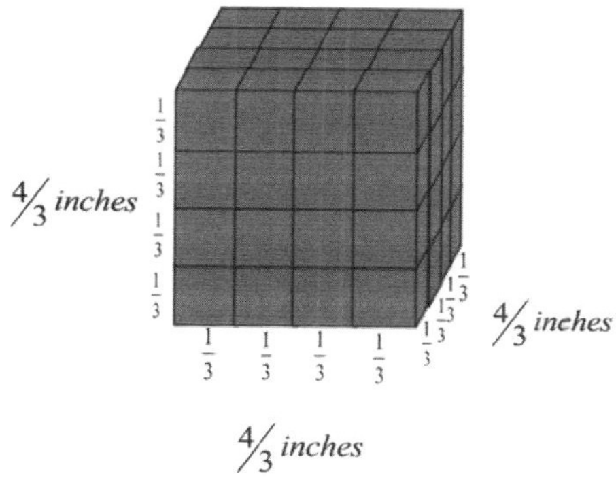

The top row of the box shows there are 16 cubes with sides of $\frac{1}{3}\,in$. There are 4 of these rows, which makes a total of $16 \times 4 = 64$ cubes. Since the volume of each little cube is $\frac{1}{3} \times \frac{1}{3} \times \frac{1}{3} = \frac{1}{27}\,in^3$, then the total volume is $64 \times \frac{1}{27} = \frac{64}{27} = 2\frac{10}{27}\,in^3$. This total volume can also be found by multiplying the dimensions of the raisin box package directly: $\frac{4}{3} \times \frac{4}{3} \times \frac{4}{3} = \frac{64}{27} = 2\frac{10}{27}\,in^3$.

EXAMPLE PROBLEM 2

At the grocery store you see two different kinds of cereal on sale for the same price. Box A has dimensions $\frac{3}{4}\,ft \times \frac{1}{2}\,ft \times \frac{2}{3}\,ft$ and Box B has dimensions $\frac{9}{10}\,ft \times \frac{1}{3}\,ft \times \frac{4}{5}\,ft$. Decide which box will give you the most cereal for your money.

Box A will give you more cereal for your money because the box can hold more volume than Box B. Find the volume of each box using the formula $V = l \times w \times h$. The volume of Box A is $\frac{3}{4} \times \frac{1}{2} \times \frac{2}{3} = \frac{6}{24} = \frac{1}{4}\,\text{ft}^3$, and the volume of Box B is $\frac{9}{10} \times \frac{1}{3} \times \frac{4}{5} = \frac{36}{150} = \frac{6}{25}\,\text{ft}^3$. Because $\frac{1}{4} = .25$ and $\frac{6}{25} = .24$, the volume of Box A is slightly larger than the volume of Box B and therefore a better deal because it will hold more cereal.

MEAN, MEDIAN, AND MODE

EXAMPLE PROBLEM 1

Define the following terms and find each for the data listed below:

Mean
Median
Mode

22, 17, 14, 15, 11, 20, 16, 15, 12, 14, 14

The *Mean* is the average of a set of data. This is found by dividing the sum of the data values by the number of values there are. The mean of the given data is 15.455, which is found by dividing the sum of the data by 11 because there are 11 terms in the data: $\frac{22+17+14+15+11+20+16+15+12+14+14}{11} = 15.455$.

The *Median* is the middle number in a set of data, and can be found by listing the data in order from least to greatest and crossing one off on either end until the middle is reached. If there are two middle terms the average of them is the median of the data. The median of the given data is 15:
$11, 12, 14, 14, 14, 15, 15, 16, 17, 20, 22$.

The *mode* is the value in a set of data that occurs the most often. In the given data the mode is 14, because that value occurs 3 times while all other values are only in the list 1 or 2 times.

EXAMPLE PROBLEM 2

The following are the test scores Michaela earned in math class. Her teacher is deciding whether the mean or the median score would best represent Michaela's final grade in math. Based on the data, determine if the teacher should use the mean or the median.

Michaela's Test Scores: 50, 55, 58, 58, 90, 92, 99

Michaela's teacher should use the mean test score. The mean test score is found by dividing the sum of the data values by how many there are. The mean of Michaela's test scores is $\frac{50+55+58+58+90+92+99}{7} = \frac{502}{7} = 71.7$. The median of Michaela's test scores is the number in the middle, which is 58: $50, 55, 58, 58, 90, 92, 99$. Since Michaela appears to have improved, it is more reasonable for her to have a passing grade of 71.7% than a failing grade at 58%. Therefore the mean better represents this data.

EXAMPLE PROBLEM 3

You want to go to the ballpark to buy tickets to a baseball game. To ensure that you are able to get tickets you will need to get in the line that has the least waiting time. At the ballpark there are two lines: Red and Blue. You surveyed your friends who have bought tickets in the past to find their wait times. Analyze the data you collected below, decide which line you should wait in to purchase tickets. Time (in minutes):

Red Line: 5, 12, 7, 15, 11, 12, 6, 10, 9 Blue Line: 9, 9, 26, 6, 10, 8, 11, 10, 8

You should wait in the Blue Line. This is because the Red Line has a median wait time of 10 minutes while the Blue Line has a median wait time of 9 minutes. This is found by first listing the data points in order from least to greatest and finding the middle number, which is the median of the data. Crossing off the numbers on the left and right side until reaching the middle will find the middle number: Red Line: $5, 6, 7, 9, 10, 11, 12, 12, 15$ and Blue Line: $6, 8, 8, 9, 9, 10, 10, 11, 26$.

EXAMPLE PROBLEM 4

You have gathered data on the wait times for two different lines at the ballpark. You decide to wait in the Blue Line because it has the lower median wait time. Your friend tells you, however, that you should wait in the Red Line because it has the shortest wait time recorded. Consider the data below, decide if you should change your mind and take your friend's advice.

Time (in minutes):

Red Line: 5, 12, 7, 15, 11, 12, 6, 10, 9 Blue Line: 9, 9, 26, 6, 10, 8, 11, 10, 8

You should not take your friend's advice. You should still choose the Blue Line because the Blue Line's median is 9 minutes, while the Red Line's median is 10 minutes. The data point of 26 in the Blue Line is considered an outlier because it varies greatly from the median compared to the other data points in the set. If you ignore that outlier, the median is still 9! Even ignoring the highest wait time for each line, the mean wait time is higher for the red line, so although the Red Line had the shortest wait time and Blue Line had the longest wait time that is not the best judge of predicting wait time.

EXAMPLE PROBLEM 5

Predict what the median of the data is by analyzing the dot plot below. Calculate the median to check the prediction.

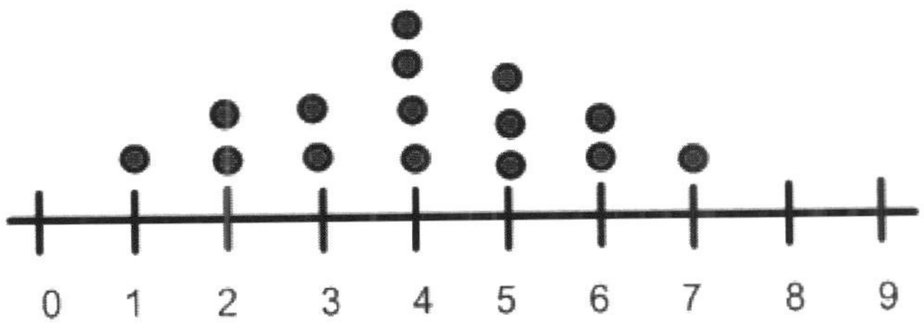

It appears that the median of the data is 4, as its values fall in the middle of all of the dots on the dot plot. Calculating the median is done by first listing the data in order from least to greatest. Each dot on the dot plot represents the number it is above, so therefore the data values are 1, 2, 2, 3, 3, 4, 4, 4, 4, 5, 5, 5, 6, 6, 7. To find the median the values must be crossed off until the middle number is reached: 1, 2, 2, 3, 3, 4, 4, 4, 4, 5, 5, 5, 6, 6, 7. The appearance of the median in the dot plot is accurate, as the median is 4.

STATISTICAL QUESTIONS AND VARIABILITY

Determine whether the following questions are considered statistical questions and explain why or why not:

'How many pets do the families in my neighborhood have?'

'What size are my shoes?'

'How many pages did I read in the last hour?'

'How long did it take the students in my class to get to school today?'

Yes, this is a statistical question because it will have variability. The families in your neighborhood will not each have the same number of pets, but there will be a variation in the data collected and conclusions can be drawn from this.

No, this is not a statistical question. There will be no variation because there is one answer.

No, this is not a statistical question. There will be no variation because there is one answer.

Yes, this is a statistical question because it has variability. There will be many different answers to this question and conclusions can be drawn from the data.

DATA REPRESENTATION AND ANALYSIS

EXAMPLE PROBLEM 1

Represent the following data in a dot plot and find the mode of the data:

The amount of money your friends make babysitting per hour

5, 2, 7, 9, 4, 5, 6, 7, 9, 7

The data is represented in a dot plot below:

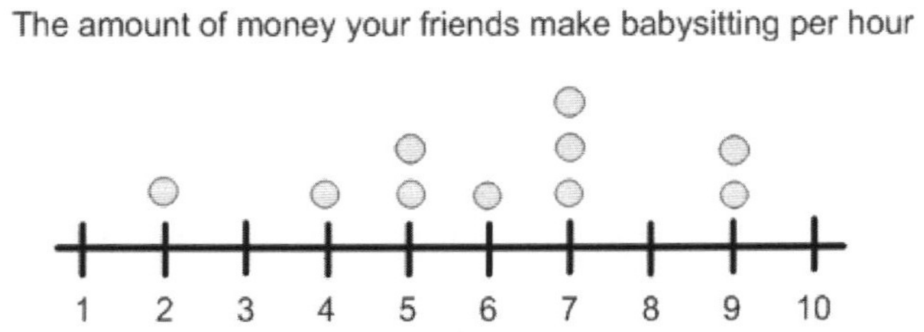

This is made by creating a number line that can display the range of data and then placing one dot above each number for each data value equal to that number. The mode of this data is 7, which is the number that occurs in the data most. The mode is easy to see in a dot plot because it is the number that has the most dots. In this data, three people get $7/hour babysitting while only one or two people get paid the other amounts in the list.

EXAMPLE PROBLEM 2

Represent the following data in a box plot:

Michael's math test grades last semester

88, 90, 95, 82, 98, 90, 77, 89, 91

The data is represented in the box plot below:

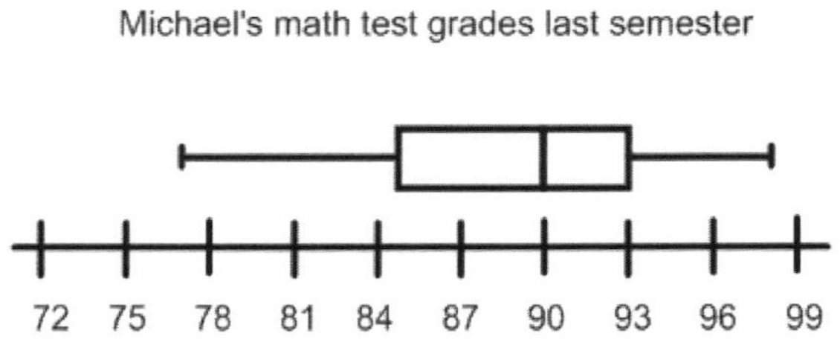

This is made by creating a number line that will fit the distribution of the data and then determining the five points that are needed to create the box plot, which are: the minimum, the maximum, the median, the lower quartile, and the upper quartile. The minimum is the number in the list with the least value, 77, and the maximum is the number with the greatest value, 98. A small tick mark is placed above those two values on the number line. The median of the data is 90, which is found by listing the data in order from least to greatest and finding the number in the middle: 77, 82, 88, 89, 90, 90, 91, 95, 98. The lower quartile is the median of the lower half of the data, which is 85 because the middle of the lower half of the data is between the two middle numbers 82 and 88, and is therefore found by finding their average: $\frac{82+88}{2} = 85$. The upper quartile is the median of the upper half of the data, which is 93 because it is between 91 and 95, and their average is $\frac{91+95}{2} = 93$. A tick mark is placed above the median and upper and lower quartile values on the number line and a box is created around those three marks. A line is then extended from the ends of the box to the minimum and maximum.

Example Problem 3

Represent the data below in a histogram:

The amount of homework your friends did last night, in minutes

45, 25, 33, 65, 37, 55, 45, 42, 31, 49, 48, 22

The histogram below represents the data:

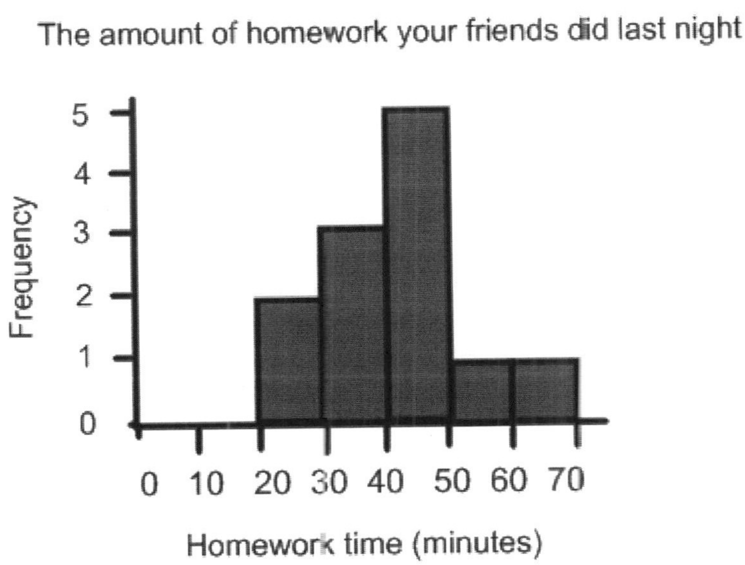

This is made by creating a graph and labeling the x-axis with homework time, the y-axis with frequency, and creating appropriate intervals for each. (Generally, divide the range of the data into 5 - 7 intervals.) The data values in each interval, which are 10 minute intervals in this histogram, are counted to determine the frequency of data values in those intervals. Adjacent rectangles are created to show the distribution of data. There are 2 pieces of data that fall between 20 and 30 minutes, so that interval has a frequency of 2. There are 3 pieces of data that fall between 30 and 40 minutes, so that interval has a frequency of 3. There are 5 pieces of data that fall between 40 and 50 minutes, so that interval has a frequency of 5. And there is 1 piece of data each that falls between 50 and 60 minutes and 60 and 70 minutes, so those each have a frequency of one.

Example Problem 4

Mr. Smith was deciding which of two intersections to post signs at for his garage sale. He sat at one intersection each day for 30 minute time intervals for two consecutive days. Summarize the meaning of his data recorded below and predict what conclusions he may draw

Number of cars that drove by

Time	Intersection 1	Intersection 2
7am-7:30	8	1
9am-9:30	10	6
11am-11:30	11	10
1pm-1:30	0	9
3pm-3:30	5	10
5pm-5:30	2	4

Mr. Smith will most likely want to place signs for his garage sale at Intersection 2. This is because over the course of the day that Mr. Smith observed Intersection 2, 40 cars drove by, whereas over the course of the day that he observed Intersection 1, only 36 cars drove by. However, Mr. Smith's data shows observations that might make him choose Intersection 1. For example, Mr. Smith might know that morning hours are usually the busiest for garage sales and thus he might want to put his signs at Intersection 1 because more cars drive through that intersection in the morning hours. But, it seems as though the traffic at Intersection 1 tapers off greatly in the afternoon and Mr. Smith might be more concerned with having the possibility of a steady flow of customers at his garage sale all day.

Example Problem 5

The local hospital recorded the lengths of the babies born yesterday, given below. Analyze the data and determine what units were being used in the measurements. With the conclusion justify any changes you would recommend for the collection of the data.

Lengths of Babies:

1.5, 1.75, 2, 16, 17.5, 17.5, 18.5, 19, 20, 20, 21, 22

It is clear that there are two different units of measurement being used to measure the lengths of babies. All of the data points between 16 and 22 are reasonable if measured in inches. The data points of 1.5, 1.75, and 2 cannot have been measured in inches and were most likely measured in feet. The data points of 1.5, 1.75, and 2 are reasonable if measured in feet, however all the other pieces of data are not reasonable to have been measured in feet. Therefore, the hospital will need to choose one unit of measure when collecting the data on the lengths of the babies born there. One suggestion would be to measure all babies in inches, and thus change the values of 1.5, 1,75, and 2 in the data to 18, 21, and 24.

Example Problem 6

Mr. Gordon is Principal at Detroit's largest Middle School and decided to allow students to choose what time they would eat lunch on a given day. To do this he surveyed a group of students and the data is represented in the histogram below. Principal Gordon concluded that he would allow students to eat at their desired lunch time, 11:00-11:30, because the most students wanted to eat during that time. Explain whether or not Principal Gordon's conclusion was justified based on the data in the histogram.

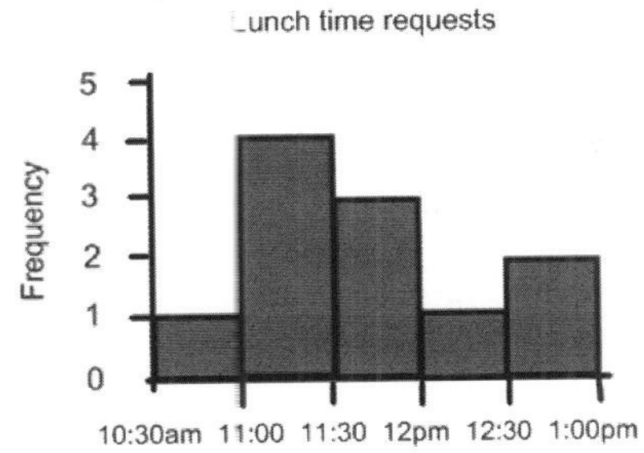

No, Mr. Gordon's conclusion is not justified based on the data given in the histogram. Although the majority of the students he surveyed do want to eat lunch between 11 and 11:30, he only surveyed 11 students. This can be found by adding the frequency of each lunch time period. Surveying a group of 11 students might be a reasonable choice for a very small school, but since Mr. Gordon is the Principal at the largest middle school in Detroit, a sample size of 11 students is not enough to draw justifiable conclusions from. It is possible that the majority of all students will want to each lunch between 11 and 11:30, but Mr. Gordon will need a larger sample size to be able to justify the conclusion that it represents the majority of students' preferences.

NET FIGURE

EXAMPLE PROBLEM 1

Sheila is going to make her own canvas tent for camping. She wants to make a tent that is the shape of a triangular prism. Given the dimensions below, create a net figure of the tent and determine how much canvas Sheila will need to make it.

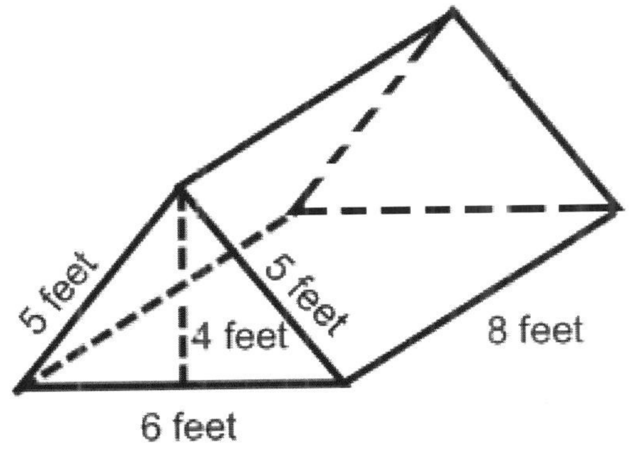

Sheila will need 152 square feet of canvas. This is found by determining the surface area of the triangular prism. In the net figure below, it is clear to see that the sum of the areas of two triangles and three rectangles will give the full surface area.

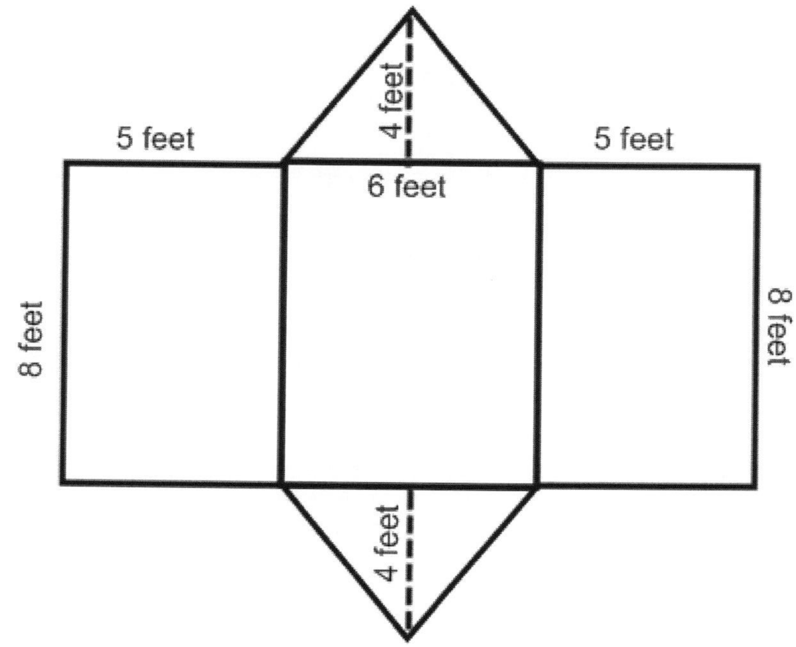

Each triangle has a base of 6 feet and height of 4 feet, so the area of each triangle is $\frac{6\times4}{2}=\frac{24}{2}=12\text{ft}^2$. Each outer rectangle in the net figure has a length of 8 feet and a width of 5 feet, so the area of each of those rectangles is $5\times8=40\text{ft}^2$. The other rectangle has a length of 8 and a width of 6, so has an area of $6\times8=48\text{ft}^2$. The total surface area is found by adding the area of all 5 of the separate shapes together: $12+12+40+40+48=152\text{ft}^2$.

EXAMPLE PROBLEM 2

Draw a net figure of the present below and determine how much wrapping paper will be needed to wrap it.

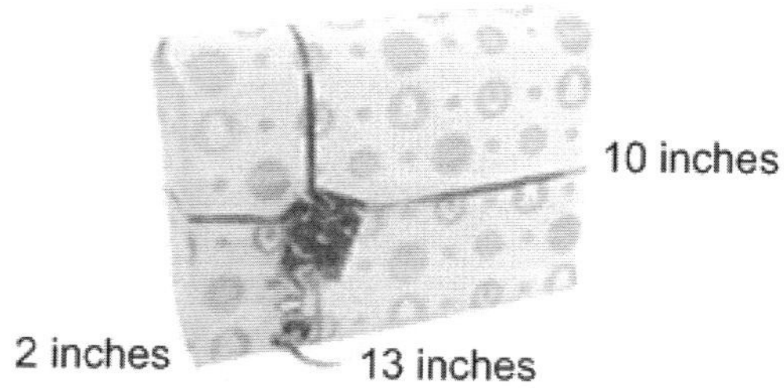

352 square inches of wrapping paper will be needed to wrap the present, which is the surface area of the box. In the net figure below, it is seen that there are two rectangles with dimensions 13in × 2in which each have an area of $13\times2=26\text{in}^2$, two rectangles with dimensions 10in × 2in which each have an area of $10\times2=20\text{in}^2$, and two rectangles with dimensions 13in × 10in which each have an area of

$13 \times 10 = 130\text{in}^2$. The total surface area is found by adding together the areas of the 6 rectangles: $26 + 26 + 20 + 20 + 130 + 130 = 352\text{in}^2$.

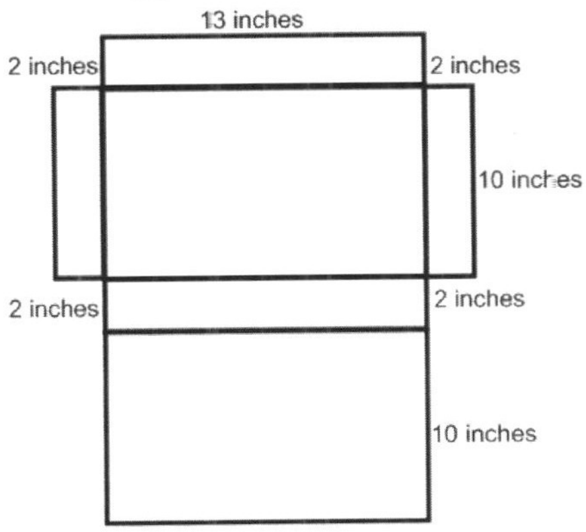

INEQUALITIES

EXAMPLE PROBLEM 1

Write inequalities to represent the following situations, and graph them on number lines:

The speed limit on the highway is 65mph.

Your teacher is older than 30 years old.

A. This is represented by the inequality $x \leq 65$, because a person can drive at any speed less than or equal to 65mph. This is graphed on the number line below, showing a closed circle on the 65 and an arrow shading to the left, because the safe speeds include 65 and anything less than it:

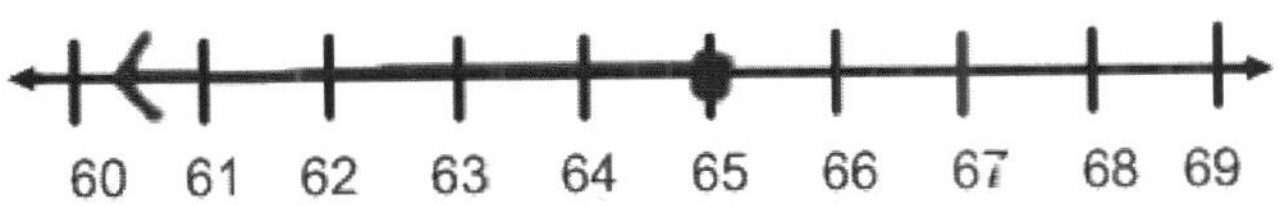

B. This is represented by the inequality $x > 30$, because your teacher is any age greater than 30 years old, but not equal to 30. This is graphed on the number line below, showing an open circle at 30 and an arrow shading to the right, which means that 30 is not included in the possible ages for your teacher but any number greater is:

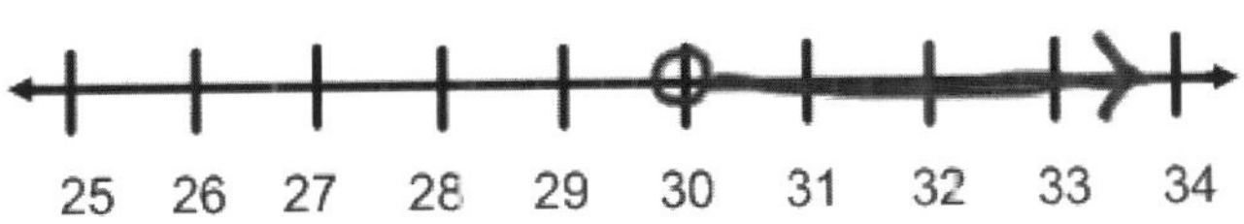

EXAMPLE PROBLEM 2

Write inequalities to represent the following situations, and give one possible solution for each:

You will not go to school for more than 5 days this week.

Your dad makes at least \$100 a day at work.

The bowling alley requires a birthday party size of less than 20 kids.

> This is represented by the inequality $x \leq 5$, because you might go to school 5 days or you might go to school any amount less than that. One solution could be 4, because you might get sick one day this week and only go to school 4 days.
>
> This is represented by the inequality $x \geq 100$, because your dad will make \$100 or more at work each day. One solution could be \$125.
>
> This is represented by the inequality $x < 20$, because there cannot be 20 kids in the party but there can be any amount less than that. One solution could be 15 kids at the birthday party.

Example Problem 3

Write an inequality to represent each of the following situations and explain the meaning:

Jim's golf score was 8 under par while Tom's golf score was 2 under par.

> *Water boils at* $212°F$ *and freezes at* $32°F$.
>
> *Yesterday the temperature was* $-17°F$ *and today the temperature is* $-21°F$.
>
> If we assume 'par' is like zero on the number line, this can be represented by $-8 < -2$, which means that it took Tom a greater number of strokes to finish his golf game, because -2 is greater than -8.
>
> This can be represented by $32° < 212°$, which means that the boiling point of water is warmer than the freezing point of water, because $212°$ is greater than $32°$.
>
> This can be represented by $-21° < -17°$, which means that it was warmer out yesterday because $-17°$ is greater than $-21°$.

Example Problem 4

Describe scenarios that might correspond to the following inequalities:

$$-20°F < -5°F$$
$$-\$50 < -\$40$$
$$-6 < -2$$

> This represents two different temperatures. One scenario could be that one day it was $-20°F$ in Alaska while the next day it was $-5°F$ there. Since $-5°F$ is greater than $-20°F$, it was warmer the second day.
>
> This represents two amounts of money. One scenario could be that Jenny owes the bank \$50 while her friend Susan owes the bank \$40. Jenny has a greater debt so therefore a more negative balance, and less money than Susan.
>
> This simply compares two negative integers. One scenario could be that they represent the scores of two golfers. If Max got a golf score of -6 ("six under par") and his friend got a golf score of -2 ("two under par"), then that means it took Max fewer swings to finish the game than it took his friend.

Example Problem 5

Represent each scenario below with an absolute value inequality and give one possible solution:

A. Last week Mr. Anderson lost 6 of his cattle to Mad Cow Disease and this week he lost even more.

B. Yesterday Andrus took an elevator from his city street underground 4 stories to the subway. Today Andrus took the same elevator down a fewer number of stories.

C. On Monday Sally withdrew $50 from her bank account and on Tuesday she made a larger withdrawal.

The inequality $|-6| < |x|$ represents the number of cattle Mr. Anderson lost this week, x, compared to the number he lost last week. One possible solution would be $x = -10$, or losing 10 cattle, because $-6| < |-10|$.

The inequality $|x| < |-4|$ represents how far below ground level Andrus traveled today, x, compared to yesterday. One possible solution would be $x = -1$, or traveling 1 story below ground level, because $|-1| < |-4|$.

The inequality $|-50| < |x|$ represents the amount of Sally's withdrawal on Tuesday x, compared to her withdrawal Monday. One possible solution would be $x = -100$ or withdrawing $100, because $|-50| < |-100|$.

Example Problem 6

Describe a scenario that the following absolute value inequalities could represent:

A. $|-10\ feet| > |-3\ feet|$

B. $|-300\ points| > |-50\ points|$

This inequality could represent the depth of two fish swimming in the ocean. One fish is 10 feet below sea level, or -10 feet, while the other fish is 3 feet below sea level, or -3 feet. The absolute value of -10 is greater than the absolute value of -3, which illustrates the fact the fish swimming 10 feet below sea level is a greater distance from sea level.

This inequality could represent changes in the value of the stock market on two different days. On one day the stock market fell 300 points and the next day the stock market fell 50 points. The absolute value of -300 is greater than the absolute value of -50, illustrating that 300 points is a greater fall than 50 points.

Example Problem 7

Write inequalities to represent the following situations, and graph them on number lines:

The minimum cost of a school photo package is $10.

A person has to be less than 48 inches tall to ride the kiddie rides at the amusement park.

This statement is represented by the inequality $x \geq 10$, because you can spend $10 or more on photos. This is graphed on the number line below, showing a closed circle on the 10 and an arrow shading to the right, because the amount you can spend includes 10 dollars or any number greater than that.

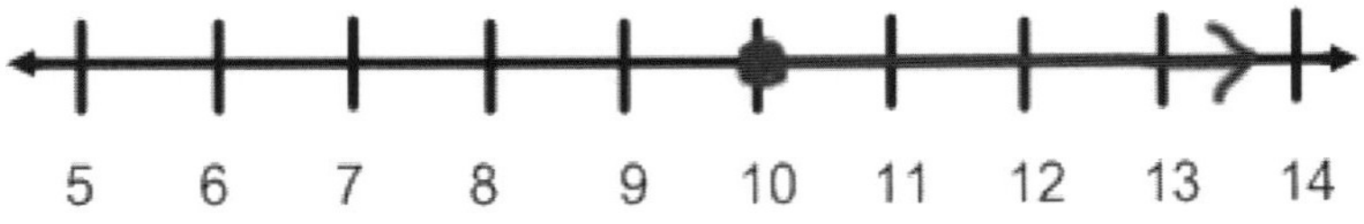

This statement is represented by the inequality $x < 48$ because the acceptable heights are all those less than 48 inches. This is graphed on the number line below, showing an open circle on the 48 and an arrow shading to the left, because 48 is not included in the acceptable heights, but anything less than it is:

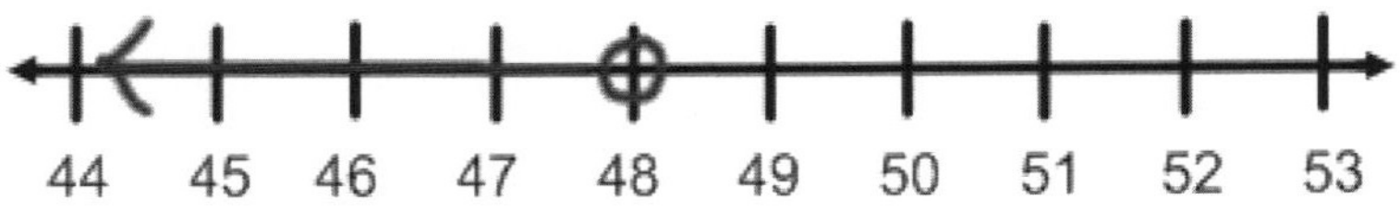

Independent Variable and Dependent Variable

The independent variable is that quantity in a situation that affects the value of the other when changed.

The dependent variable is that quantity in a situation that depends on the value of another variable.

For the following examples:

The total amount of money m you earn for mowing n lawns.

The price p you pay for renting x videos.

The distance d you travel on a bus over time t.

The number of lawns you mow, *n*, is the independent variable, while the amount of money you earn, *m*, will depend on how many lawns you mow and thus is the dependent variable.

The number of videos, *x*, you rent is the independent variable, and the price, *p*, you pay is the dependent variable because it depends on how many videos you rent.

The time, *t*, is the independent variable, and the distance, *d*, is the dependent variable because the distance you travel will depend on how long you travel for.

Opposite

The 'opposite' of a number is the number that is the same distance away from zero as the given number but on the other side of zero on a number line. The 'opposite' of a number always has a sign opposite the given number.

The opposite of -4 is 4.

The opposite of 7 is -7.

The opposite of 21 is -21.

The opposite of -10 is 10.

The opposite of 0 is 0. This is because the only number that is the same distance zero is from zero is zero. Zero is neither positive nor negative, so it will not change signs.

The opposite of 13 is -13.

The opposite of -13 is 13.

NUMBER LINE

EXAMPLE PROBLEM 1

Label a number line with the following values:

Point A: 2

Point B: The opposite of 3

Point C: 0

Point D: -2

Point E: The opposite of -5

Point F: The opposite of -1

The number line below shows the points:

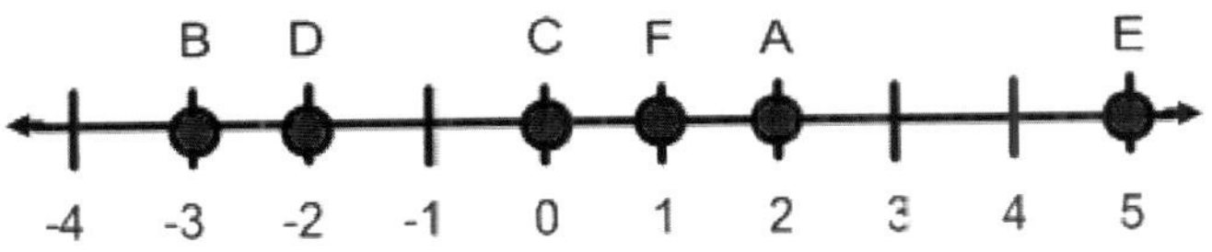

Point A is located at 2 on the number line. Point B is located at -3, which is the opposite of 3. Point C is located at 0. Point D is located at -2. Point E is located at 5, which is the opposite of -5. And Point F is located at 1, which is the opposite of -1.

EXAMPLE PROBLEM 2

Determine the following values, then plot the points on a number line and write an inequality that represents each relation between the points.

5 units to the right of -2

3 units to the left of 1

The number that is 5 units to the right of -2 is 3. The number line below shows these values and it is seen that $3 > -2$ because 3 is located to the right of -2.

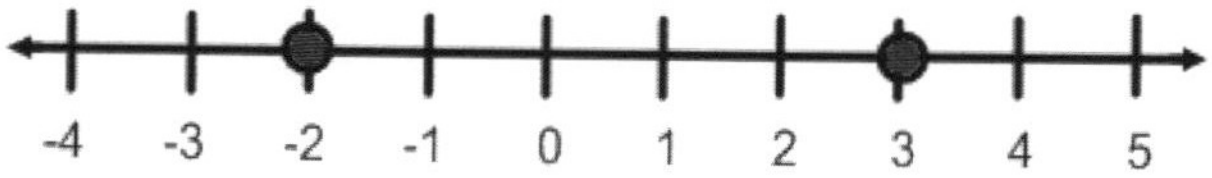

The number that is 3 units to the left of 1 is -2. The number line below shows these values and it is seen that $-2 < 1$ because -2 is located to the left of 1.

Example Problem 3

Determine the following values, then plot the points on a number line and write an inequality that represents each relation between the points:

1 unit to the left of 1

2 units to the left of -2

A. The number that is 1 unit to the left of 1 is 0. The number line below shows these values and it is seen that $0 < 1$ because 0 is located to the left of 1.

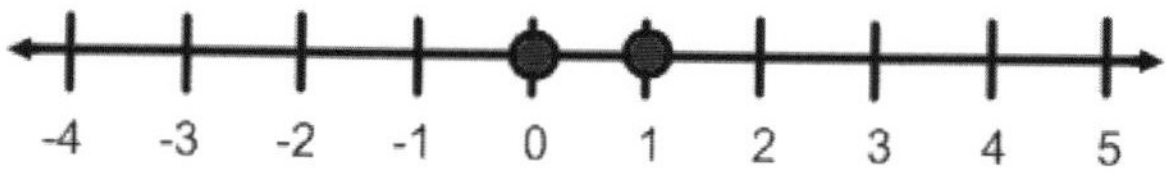

B. The number that is 2 units to the left of -2 is -4. The number line below shows these values and it is seen that $-4 < -2$ because -4 is located to the left of -2.

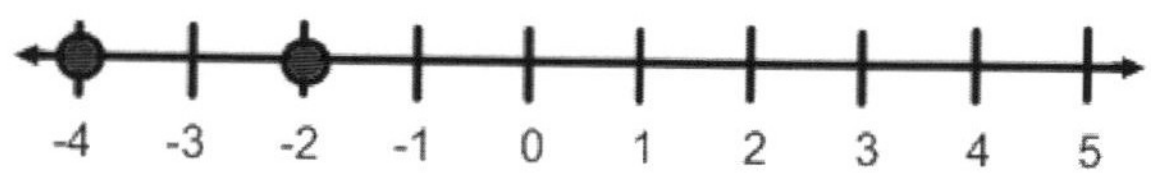

Absolute Value

Absolute Value is how far away a number is from zero on a number line, or the distance between zero and that number. The absolute value of a number is always positive, because distance is positive.

The absolute value of 40 is 40, because it is 40 units away from zero.

The absolute value of -12 is 12 because it is 12 units away from zero.

The absolute value of -25 is 25 because it is 25 units away from zero.

The absolute value of 18 is 18 because it is 18 units away from zero.

The absolute value of -100 is 100 because it is 100 units away from zero.

EXAMPLE PROBLEM 1

Write an absolute value equation to express the following situations:

After the first day of a bike race, Peter is 45 seconds behind the lead racer.

Marla owes her mom $10.

The scuba diver is 20 feet below sea level.

This situation can be represented by $|-45| = 45$. When we say a racer is behind, it actually means that racer's time is longer than the leader's. To catch the leader, then, Peter needs to subtract 45 seconds from his time, and this equation shows that the difference between the times is 45 seconds.

This situation can be represented by $|-10| = 10$. This shows that Marla is in debt $10 to her mom, but Marla does not actually have a negative amount of money so it is expressed as absolute value, which shows that Marla needs to pay $10 to her mom.

This situation can be represented by $|-20| = 20$. This shows that the scuba diver is 20 feet under the measurement we mark as '0' on the earth, sea level. However, the scuba diver is not actually a negative distance from sea level so it is expressed in absolute value, showing the scuba diver is a distance of 20 feet away from sea level.

RANGE

The range helps to quantify the amount of variation in a set of data. It is the difference between the maximum value and the minimum value and can be found by subtracting the minimum value in a set from the maximum value in a set.

EXAMPLE PROBLEM 1

Find the range of each situation below and describe its meaning:

The height of students in Mickey's class (in inches):

58, 62, 59, 61.5, 61, 58, 60, 59.5, 61, 58, 59, 61, 60.5, 58.5

The amount of money each friend spent at dinner (in dollars):

15, 7, 11.50, 21, 14, 17.50, 10, 9, 23.50, 14

For the height of students in Mickey's class the range is 4 inches. The maximum height is 62 inches and the minimum height is 58 inches, so therefore the range is $62 - 58 = 4$. This means that all students in Mickey's class are fairly close together in height as this is a small range to hold all of the data. If the whole class was lined up the difference in height from the tallest person to the shortest would be only 4 inches.

For the amount of money each friend spent at dinner, the range is $16.50. The maximum value is $23.50 and the minimum value is $7, so therefore the range is $23.50 - 7 = 16.50$. This means that the friends spent a wide range at dinner, meaning some friends bought much more expensive meals or more food than some of the others.

EXAMPLE PROBLEM 2

The following data shows the cost of used cars in the sales lot. Plot the data on a line plot and describe the meaning of the range of data.

Price of Used Cars For Sale (in dollars)

14,000	7,500	8,000	600
19,000	18,500	9,500	3,000
5,500	950	16,000	2,000

The data is displayed on the line plot below:

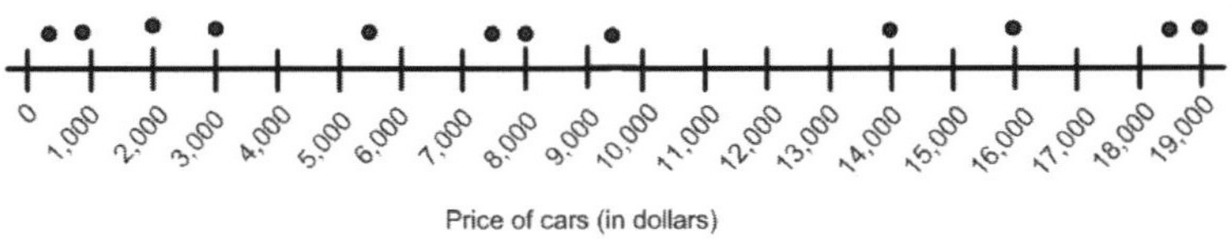

The range of the data is $18,400. This is found by subtracting the smallest value, $600, from the largest value, $19,000. This is a large range for the price of the cars in the lot. Looking at the line plot it is easy to see that the prices of cars are very spread out. This might mean that the lot has some fairly new used cars and also some very old cars for sale.

EXAMPLE PROBLEM 3

The following table shows the amount of television each student watched the night before the math test. Any student who watched television for a greater amount of time than the median earned a D on the test. Find the range of the data and determine who got a D on the test.

Student	Hours of TV watched	Student	Hours of TV watched
Billy	1	Brock	2
Paige	6	Colleen	3.5
Margaret	.5	Courtney	0
Skylar	2	Blair	3
James	0	Theresa	1
Fiona	0	Tyler	4.5
Kyle	3	Troy	1.5

The range of the data is 6 hours. This is found by subtracting the smallest value, 0, from the largest value, 6.

The median of the data must be found in order to determine who earned a D on the test. The median of the data is 1.75 hours. This is found by ordering the numbers from least to greatest and finding the middle number, like so: 0, 0, 0, .5, 1, 1, 1.5, 2, 2, 3, 3, 3.5, 4.5, 6. Since there are two numbers in the middle of the data, the average of those two must be found. Therefore, the median of the data is $\frac{1.5+2}{2} = \frac{3.5}{2} = 1.75$. Since any student who watched TV for more than 1.75 hours earned a D on the test, Paige, Skylar, Kyle, Brock, Colleen, Blair, and Tyler all earned D's on the test.

EXAMPLE PROBLEM 4

The students in Mark's class filled in the table to represent how many siblings they each have. Use the table to find the range of the data and describe the shape made by a line plot of the data.

Number of siblings	Number of students
0	II
1	IIII
2	~~IIII~~ II
3	III
4	II
5	
6	I

The range of the data is 6. This is found by subtracting the smallest number, 0, from the largest number, 6. The shape of the line plot of this data is similar to a bell curve, or that of a small hill. The majority of the dots are above the 2, which is the mode of the data because it occurs most frequently. It is also the median of the data, because it is the value in the middle. The dots on either side of 2 are similar-sized groups getting smaller the farther away from 2 the values get. This can be seen in the line plot below:

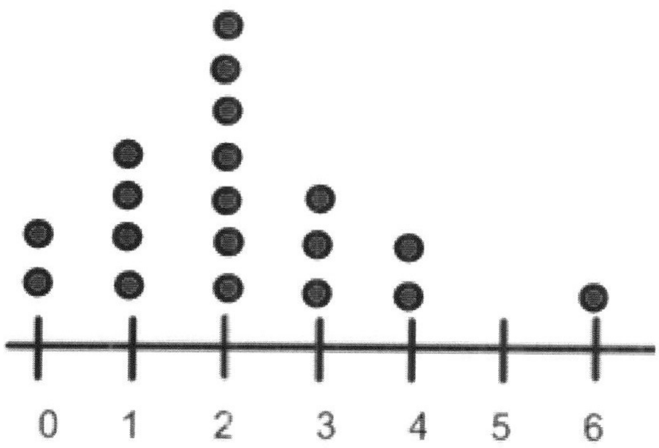

EXAMPLE PROBLEM 5

Given the following graph of data of four students' speeds to walk to school, determine what units must have been used.

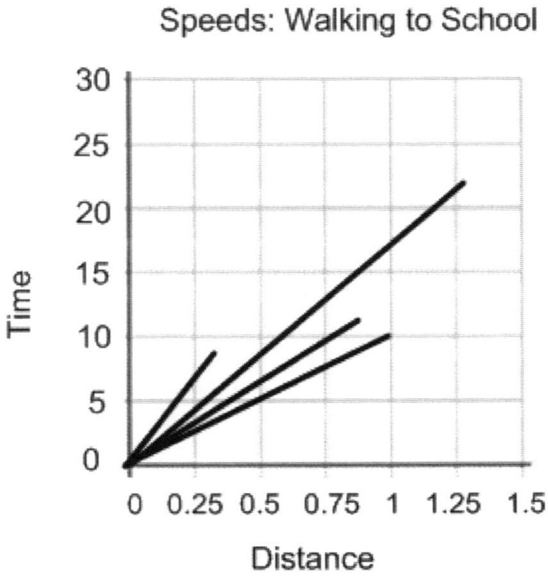

The units that must have been used are Distance in Miles and Time in Minutes. If the distance measurement is considered first, there is no other reasonable measurement other than miles. It would not make sense for the measurements to be in feet, as walking less than 1.5 feet to school would not be considered because everyone has to walk at least a few feet to get into the building, and no one can live 1.5 feet from the school. Similarly, meters or yards or anything less than those would also not be a reasonable measurement.

Considering that the distance is measured in miles, the most likely unit of time measurement is in minutes. Hours is very unreasonable as it would not take someone over 20 hours to walk 1.25 miles, and seconds is as equally unreasonable as it would not take someone just over 20 seconds to walk 1.25 miles.

The only other possible measurement would be distance in kilometers, but the times in minutes are maybe a little too slow to represent walking distances in kilometers.

IMPORTANT TERMS

Coefficient - A coefficient is a number that multiplies a variable and that denotes how many instances of that variable are represented together. This number is written in front of, or to the left of, the variable. There is generally no multiplication sign between the coefficient and the variable. For example, in the term '3x', 3 is the coefficient.

Data set - Data set is a list, or set, of all the numbers or values that are the results from a statistical experiment. An example would be that the following data set represents the shoe sizes of Fred's friends: {6, 7.5, 8, 8, 8.5, 9.5, 10}. That list of numbers is the data set and each number represents the shoe size of Fred's friends.

Difference - The result of a subtraction operation. The difference between two numbers can be thought of as the distance between them on a number line. For example, the difference of 12 and 8 is 4.

Equation - An equation is a relationship equating two expressions. An equation can be distinguished from an expression by the equal sign. For example, $4x = 12$ is an equation because it has an equal sign.

Expression - An expression is a group of one or more numbers variables, and/or terms that are combined together with operations. There is no equal sign in an expression. For example, $7x - (2x + 3) \div 4$ is an expression.

Factor - A factor is a number divides evenly into a given number For example, 4 is a factor of 8, because 8 is evenly divisible by 4.

First quartile - The first quartile is the number that is the median of the lower half of a data set that is, the half between the minimum and the median. All pieces of data less than the first quartile are in the lowest 25% of the data. It can also be referred to as the 'lower quartile'.

Greatest common factor - The greatest common factor is the greatest number that all numbers in a set are evenly divisible by.

Inequality - An inequality is a comparison between two or more expressions that are not necessarily equal; one might be greater or less than another. Because of this, when solving an inequality for a variable, the solution is a range of values. For example, $5a + 2 \leq 22$ is an inequality because it has a "less than or equal to" symbol between the two expressions.

Integer - An integer is any whole number, including 0 and all the negative whole numbers as well. The values $-100, -12, -1,\ 4,$ and 13 are all integers because they are all whole numbers, whether positive or negative. The values -2.65, $\frac{1}{3}$, and $\sqrt{6}$ are not integers because they are not whole numbers.

Least common multiple - The least common multiple is the smallest number that is evenly divisible by all numbers in a given set.

Maximum - The maximum is the greatest value in a data set.

Minimum - The minimum is the least value in a data set.

Prime factor - A prime factor is a prime number that divides evenly into another number. For example, 3 is a prime factor of 21, because 3 is a prime number and 21 is evenly divisible by 3.

Prime number - A prime number is a number that is only evenly divisible by 1 and itself.

Product - The result of a multiplication operation. For example, the product of 6 and 3 is 18.

Quotient - The result of a division operation. For example, the quotient of 10 and 5 is 2.

Rational number - A rational number is any number that can be expressed as a fraction. Equivalently, a rational number is any number that can be expressed as a terminating or repeating decimal. Rational numbers can be both positive and negative. The values $-\frac{2}{5}$ and 3 are rational numbers because $-\frac{2}{5}$ is a fraction and 3 can be expressed as a fraction as $\frac{3}{1}$. Any integer can be expressed as a fraction as itself over 1, so all integers are also rational numbers. The values $\sqrt{8}$ and π are not rational numbers (they are called 'irrational numbers') because they cannot be expressed as a fraction. These numbers are decimals that continue on without ending or repeating.

Sample size - Sample size is the number of people surveyed or the number of observations made in a statistical experiment. For example, if the mayor of a town surveyed 800 residents to determine the town's opinion on a new park, then the sample size is 800 and the mayor has hopefully judged that this will be a large enough sample to represent the whole population.

Second quartile - The second quartile is more commonly called the median, and is the middle number of a data set. It splits the data in half, so that 50% of the data is less than this number and 50% of the data is greater than this number.

Sum - The result of an addition operation. For example, the sum of 4 and 7 is 11.

Term - A term is a number or variable or a product of number(s) and/or variable(s) that does not contain an operation symbol. For example, in the expression $y + 4x + 7$, y, 4x, and 7 are separate terms.

Third quartile - The third quartile is the number that is the median of the upper half of a data set, that is, the half between the median and the maximum. All pieces of data greater than it are in the top 25% of the data. It can also be referred to as the 'upper quartile'.

Variable - A variable is an unknown value represented by a letter in an expression or equation. A variable can represent one value or a set of values. For example, in the equation $2x + 4 = 10$, x is a variable and has one value. In the equation $y = 8x + 1$, x and y can have multiple values as long as they both make the equation true.

Mathematics Practice Test #1

1. The population of New York is about twice that of Paris. Write the ratio of the population of New York to the population of Paris.

a. 1:1
b. 1:2
c. 2:½
d. 2:1

2. You pay $16 to buy four pizzas. How much did each pizza cost?

a. $1.60
b. $2.50
c. $4.00
d. $6.40

3. You can exchange US dollars for euros at the rate of 5 to 4, respectively. The table below shows the number of euros you can get in exchange for different dollar amounts.

Dollars	Euros
5	4
10	8
15	12
20	

What is the missing value in the table?

a. 16
b. 18
c. 20
d. 24

4. It takes Emily 12 minutes to run 2 miles. If she runs at the same average speed, how far will she run in 36 minutes?

a. 6 miles
b. 8 miles
c. 9 miles
d. 12 miles

5. What is 40% of 50?

a. 10
b. 16
c. 20
d. 25

6. One inch is equivalent to 2.54 cm. About how many inches are in 5 cm?

a. 0.5 in
b. 2 in
c. 2.5 in
d. 12.5 in

7. A wooden board is 3 feet long. How many pieces can the board be cut into if each piece is 3/4 of a foot long?

a. 2
b. 3
c. 4
d. 5

8. Calculate 390 ÷ 15.

a. 20
b. 26
c. 40
d. 46

9. Calculate (43 + 120) × 13.

a. 652
b. 1603
c. 1989
d. 2119

10. What is the greatest common factor (GCF) of 32 and 40?

a. 8
b. 16
c. 20
d. 160

11. What is the least common multiple (LCM) of 8 and 12?

a. 16
b. 24
c. 48
d. 96

12. You owe a friend $12. Represent your debt as a signed number.

a. −12
b. 0
c. $\frac{1}{12}$
d. 12

13. What is the additive inverse (or opposite) of 0?

a. −1
b. 0
c. $\frac{1}{2}$
d. 1

14. Graph the point (–1, 4) on a coordinate plane.

a. 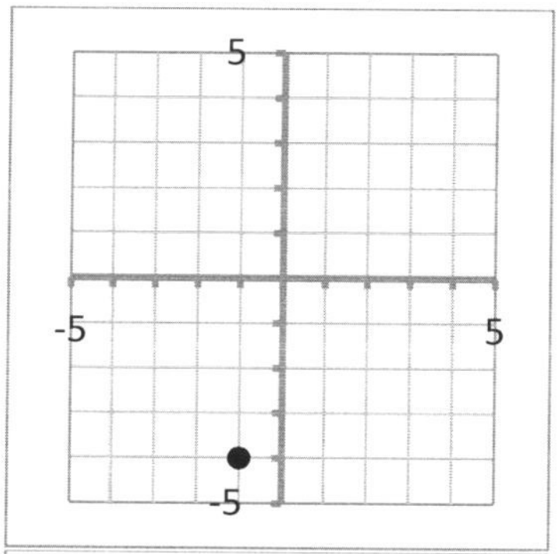

b. 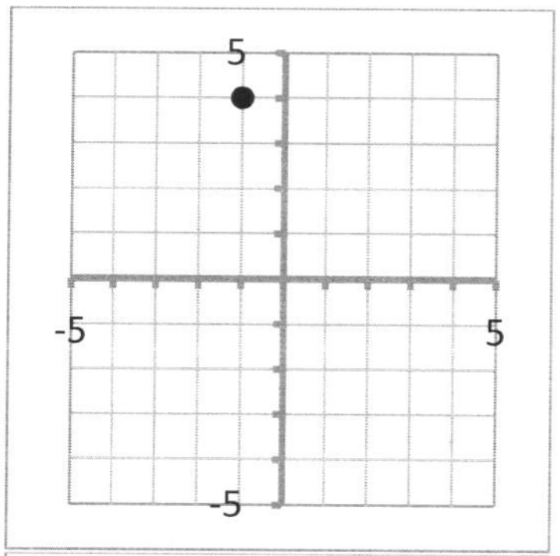

c. 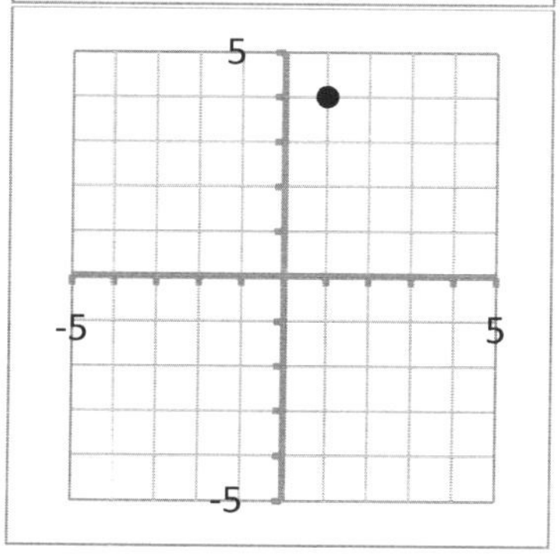

d. 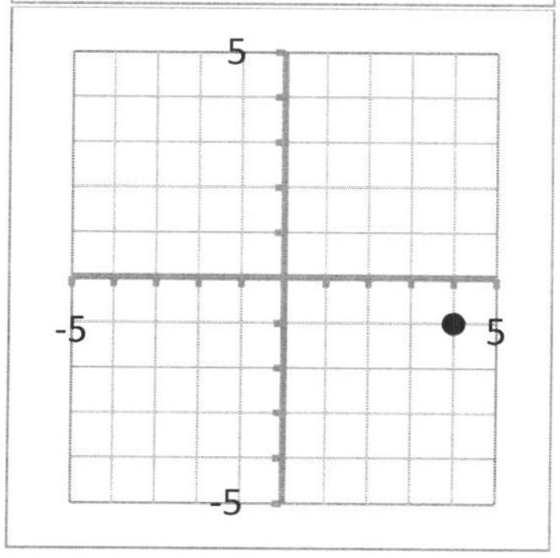

15. Graph the number –2 on a number line.

a. 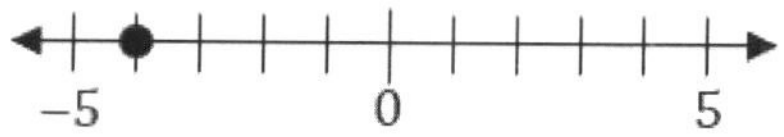

b. 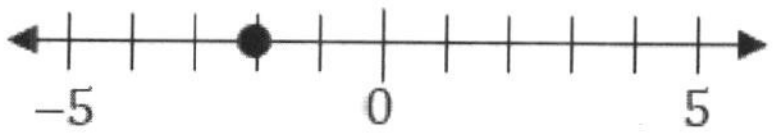

c. 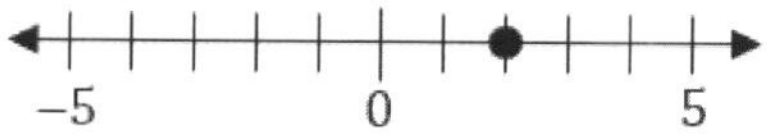

d. 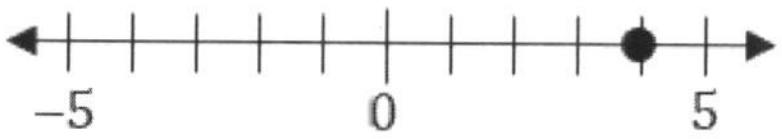

16. Which of the following is a true inequality?

a. $-5 > 2$
b. $-3 < -5$
c. $-2 > 0$
d. $-2 < -1$

17. Write an inequality that represents the statement "30 ft below sea level is lower than 20 ft above sea level."

a. $-30 < 20$
b. $0 < 50$
c. $30 < 20$
d. $30 > -20$

18. Evaluate $|-15|$.

a. −15
b. 0
c. 5
d. 15

19. The balance on your savings account is −50, which means that you have $50 of debt. A monthly charge increases the amount of your debt. Which of the following could be your new balance?

a. 60
b. 40
c. −40
d. −60

20. The vertices of a rectangle are (−3, −1), (−3, 2), (1, 2), and (1, −1). What is its perimeter?

a. 3
b. 6
c. 7
d. 14

21. Evaluate 4^3.

a. 8
b. 12
c. 64
d. 128

22. Write a mathematical expression for the sum of x and 6.

a. $x + 6$
b. $x - 6$
c. $6 - x$
d. $6x$

23. Calculate $24 \div (6 - 2)$.

a. 2
b. 3
c. 6
d. 8

24. The formula for the surface area A of a cube with sides of length s can be calculated using the formula $A = 6s^2$. Calculate the surface area of a cube with sides of length 3.

a. 27
b. 36
c. 54
d. 324

25. Simplify $3a + 2a$.

a. a
b. $5a$
c. $6a$
d. $6a^2$

26. Which expression is NOT equivalent to $4x + 2(x + 1)$?

a. $6x + 2$
b. $2(3x + 1)$
c. $3(2x + 1)$
d. $2x + 2 + 4x$

27. Which of the following is a solution to the inequality $9 + 2x > 5x$?

a. 2
b. 3
c. 4
d. 8

28. Christine drives at 60 mph down a highway. How far will she travel (in miles) after t hours of driving at this speed?

a. $t + 60$
b. $\frac{t}{60}$
c. $\frac{60}{t}$
d. $60t$

29. Solve the equation $x + 4 = 8$.

a. 2
b. 4
c. 12
d. 32

30. Graph the solutions to the inequality $x < 3$ on a number line.

a. 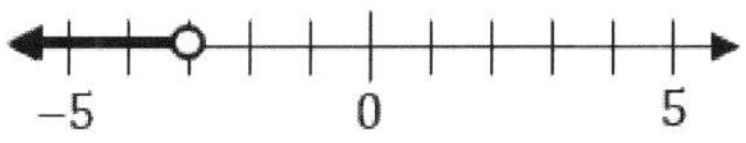

b. 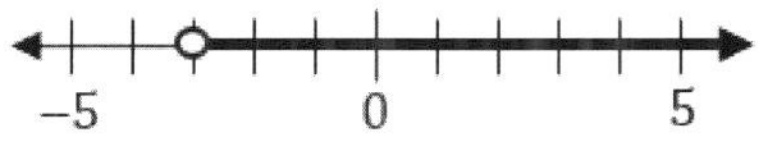

c. 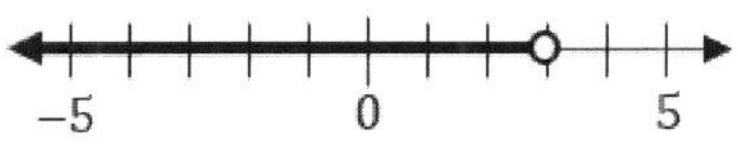

d. 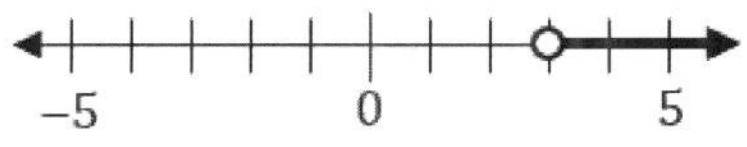

31. The dimensions of a rectangular cardboard box are $2\frac{1}{2}$ feet by $1\frac{1}{3}$ feet by 3 feet. What is its volume?

a. $6\frac{5}{6}$ cubic feet
b. 10 cubic feet
c. 12 cubic feet
d. $13\frac{2}{3}$ cubic feet

32. Bob is pumping water into his empty pool at a rate of 50 gallons per minute. Make a graph showing the amount of water in the pool during the first 10 minutes.

a. 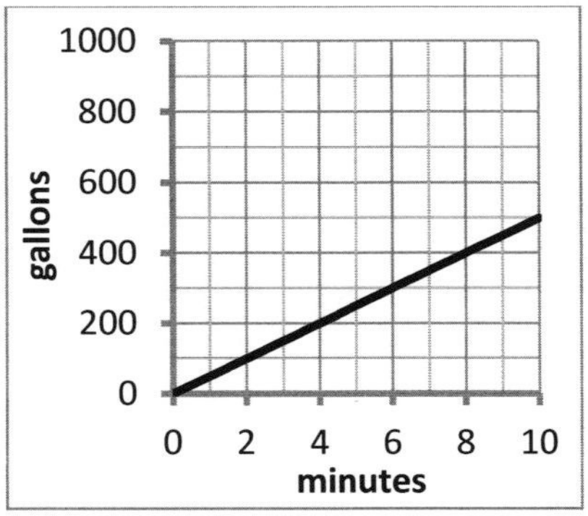

b. 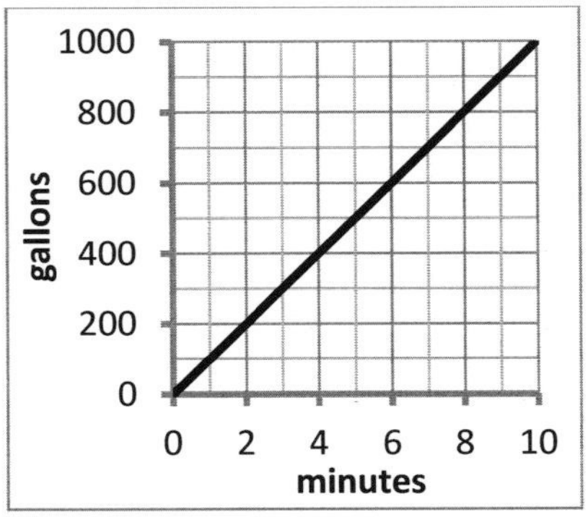

c. 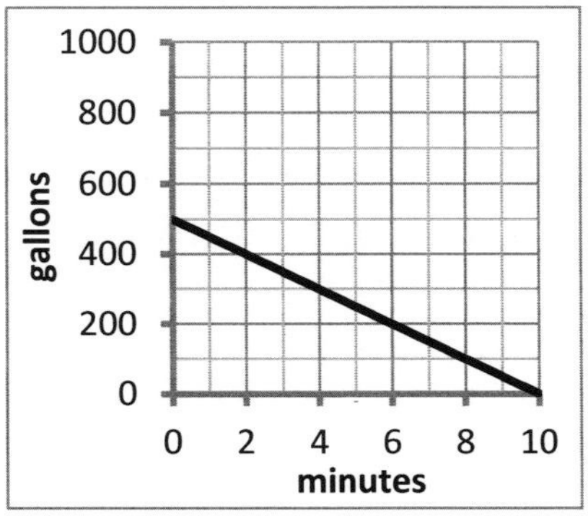

d. 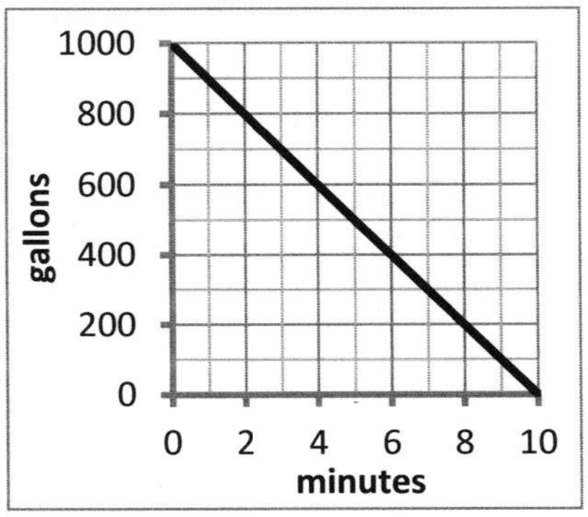

33. Calculate the area of the right triangle below.

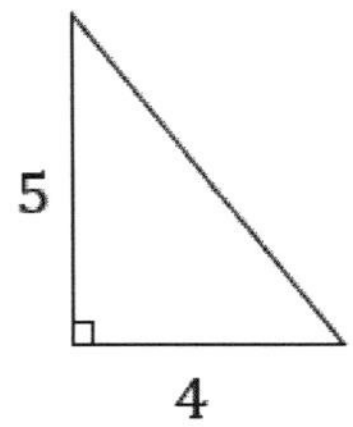

a. 9
b. 10
c. 15
d. 20

34. Charlie wants to build a kite that is 16 inches long and 21 inches wide as shown in the figure below.

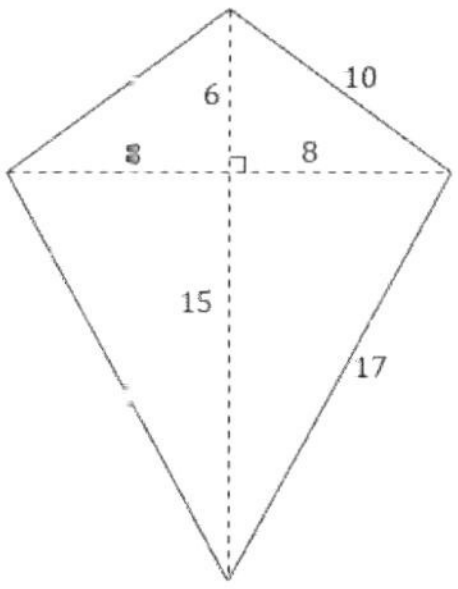

What is the area of the kite?

a. 85 square inches
b. 120 square inches
c. 168 square inches
d. 170 square inches

35. The vertices of a rectangle are (−1, 4), (3, 4), (3, 1), and (−1, 1). What is its area?

a. 6
b. 8
c. 10
d. 12

36. Which of the following is a net of a square pyramid?

a. 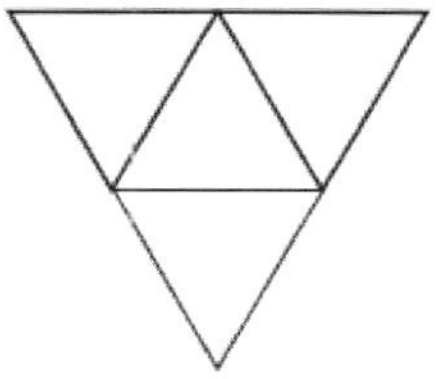

b. 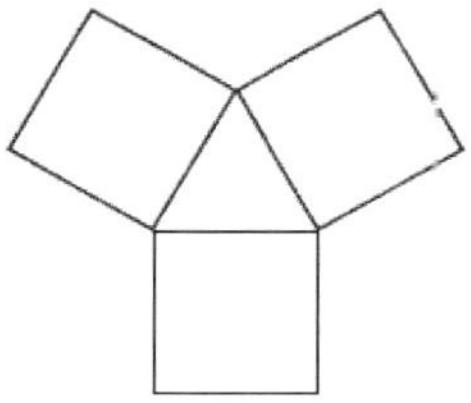

c. 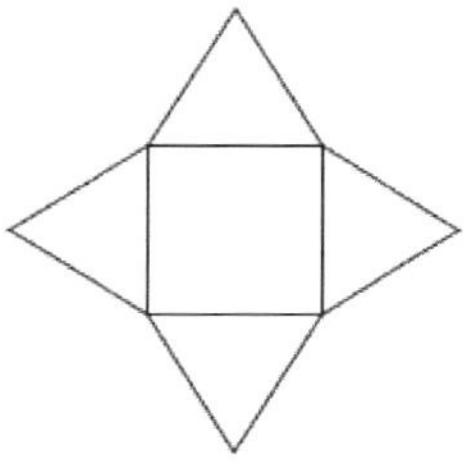

d. 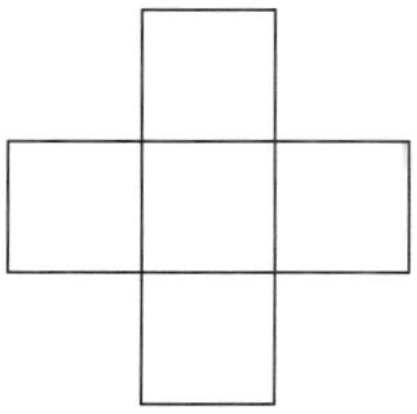

37. Which of the following is a statistical question?

a. How many students are in my English class?
b. How much money is in my bank account right now?
c. How old is Tiger Woods?
d. What are my classmates' favorite colors?

38. The mean of a data set is 12, and its range is 8. Which of the following values could NOT be value in the set?

a. 4
b. 8
c. 12
d. 16

39. Calculate the median of the data set: 8, 9, 2, 3, 4, 2, 7.

a. 2
b. 4
c. 5
d. 7

40. Create a box plot for the data set: 5, 9, 2, 8, 2, 10, 5, 5, 7, 6, 3.

a.
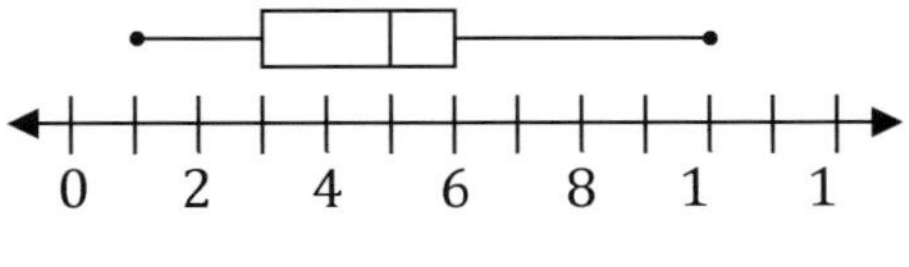

b.
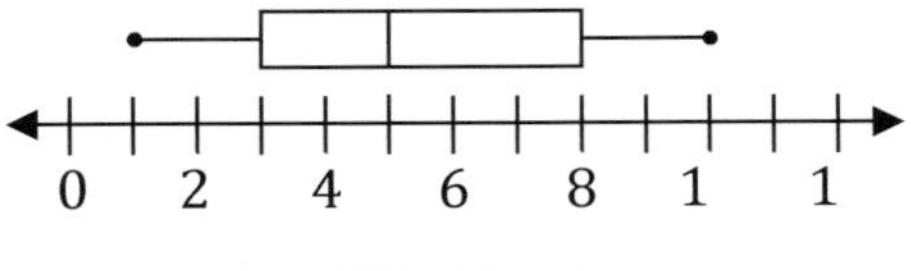

c.
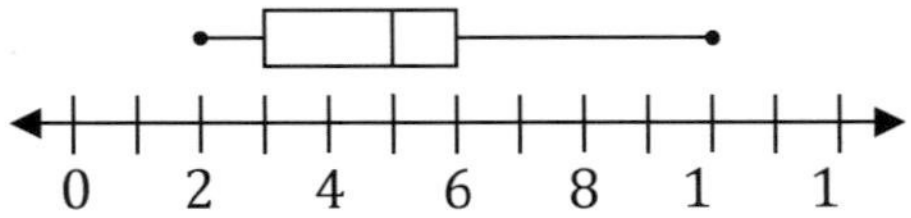

d.
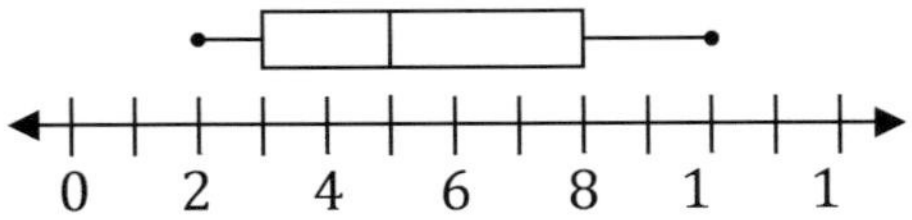

41. The test scores in Mr. Smith's class are shown below.

82, 94, 81, 70, 72, 78, 95, 83, 89, 80, 55, 77

Create a histogram to display the data.

a.
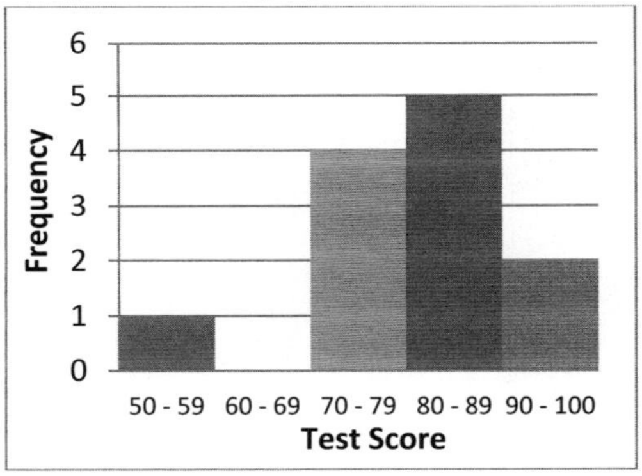

b.
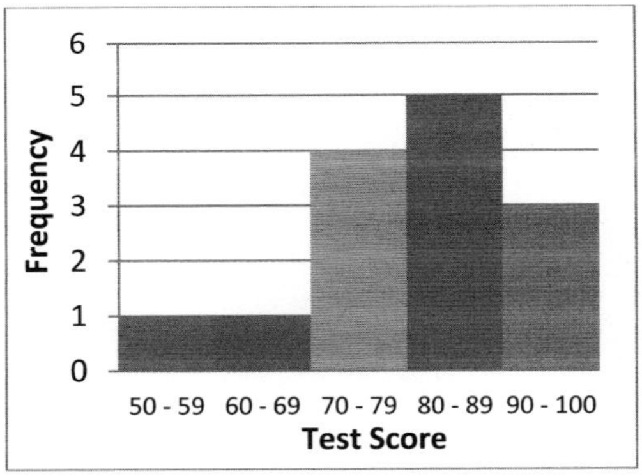

c.
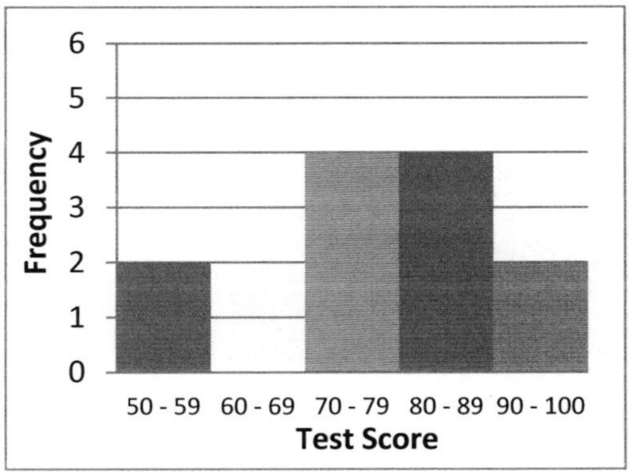

d.
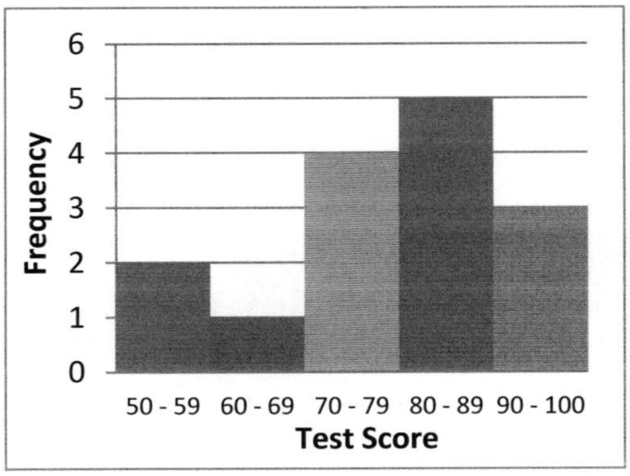

42. What unit of measure would be appropriate to use when investigating the attendance of nine different Broadway shows?

a. days
b. dollars
c. people
d. shows

43. Calculate the interquartile range of the data set: 14, 15, 11, 19, 15, 27, 16.

a. 5
b. 8
c. 15
d. 16

44. You notice that the distribution of the salary of employees in a company is skewed to the right. Which of the following would be the best choice for the measure of center of the data?

a. mean
b. median
c. mode
d. range

Answer Key and Explanations for Test #1

Math.Content.6.RP.1

1. D: A ratio is a relationship between two numbers of the same type. For instance, if the ratio of apples to oranges in a basket is 3:2, then the basket contains 3 apples for every 2 orange. In the given problem, we are looking for ratio between the populations of two cities. Since we are told that the population of New York is about twice that of Paris, there are about two New Yorkers for every Parisian. Therefore, the correct ratio is 2:1.

Math.Content.6.RP.2

2. C: The rate of total price (in dollars) per number of pizzas is 16:4. To find the unit rate (in this case, the number of dollars per pizza), write this ratio as a fraction and reduce so the denominator is equal to 1. In this case, $\frac{16}{4}$ is equal to $\frac{4}{1}$. Therefore, the unit rate is $4 per pizza.

Math.Content.6.RP.3a

3. A: The four numbers in the left column (5, 10, 15, 20) increase by 5 from top-to-bottom. The numbers in the right column (4, 8, 12) increase by 4 from top-to-bottom. Since 4 + 12 = 16, the missing value is 16. Or, each row should reduce to 5 dollars for 4 euros ($\frac{10}{8} = \frac{5}{4}$, etc.), and therefore $\frac{20}{x}$ must reduce to $\frac{5}{4}$, so that x is 16.

Math.Content.6.RP.3b

4. A: Since speed = distance ÷ time, to find Emily's average speed, divide the distance she runs by the time it takes:

$$\frac{2 \text{ miles}}{12 \text{ minutes}} = \frac{1}{6} \text{ miles per minute}$$

Therefore, her average speed is $\frac{1}{6}$ (or about 0.167) miles per minute. To determine how far she will run in 36 minutes, multiply both sides of the speed equation by time to obtain distance = speed × time:

$$\frac{1}{6} \text{miles per minute} \times 36 \text{ minutes} = 6 \text{ miles}$$

Math.Content.6.RP.3c

5. C: A percent represents one part out of 100. Thus, 40% is equivalent to 40 parts out of 100, or $\frac{40}{100}$. To find 40% of 50, multiply $\frac{40}{100}$ by 50:

$$\frac{40}{100} \times 50 = 20$$

Math.Content.6.RP.3d

6. A: We are told that the ratio of inches to centimeters is 1:2.54. To convert a length to inches, write this ratio as a fraction with inches in the numerator (i.e. on top), and then multiply the length in centimeters by that fraction:

$$\begin{aligned}\frac{1\text{ in}}{2.54\text{ cm}} \times 5\text{ cm} &= \frac{5}{2.54}\text{ in} \\ &= 1.9685\text{ in} \approx 2\text{ in}\end{aligned}$$

Math.Content.6.NS.1

7. C: To determine how many 3/4-in pieces you can cut, figure out how many times ¾ "goes into" 3, by dividing 3 by ¾:

$$3 \div \frac{3}{4}$$

To divide by a fraction, invert the fraction by exchanging its numerator and denominator, and change the division to multiplication. Then perform the multiplication normally:

$$\begin{aligned}3 - \frac{3}{4} &= 3 \times \frac{4}{3} \\ &= \frac{3}{1} \times \frac{4}{3} \\ &= \frac{4}{1} \\ &= 4\end{aligned}$$

Therefore, the board can be cut into 4 pieces.

Math.Content.6.NS.2

8. B: Perform long division to divide the dividend (in this case, 390) by the divisor (15) using a division sign called a tableau.

$$15 \overline{)\ 390}$$

Start by trying to divide the first digit in the dividend, 3, by 15. It is too small since 15 does not go into 3, so add the next digit. Dividing 39 by 15 gives you 2 with a remainder. Put the 2 at the top of the tableau above the 9, and then multiply 2 by 15 and subtract the result from 39 to find the remainder.

$$\begin{array}{rl} & \ \ 2 \\ 15 & \overline{)\ 390} \\ - & \ \ \underline{30} \\ & \ \ \ 9 \end{array}$$

Next bring down the next digit in 390 (which is a 0), and repeat the last step by dividing 90 by 15.

$$\begin{array}{r|l} & 26 \\ \hline 15 & 390 \\ - & \underline{30}\downarrow \\ & 90 \\ & -\underline{90} \\ & 0 \end{array}$$

Thus, 390 ÷ 15 = 26.

Math.Content.6.NS.3

9. D: By order of operations, perform the addition first since it is in parentheses.

$$\begin{array}{r} 43 \\ +\,\underline{120} \\ 163 \end{array}$$

Therefore, (43 + 120) × 13 = 163 × 13. Compute the product of 163 and 13.

$$\begin{array}{r} 163 \\ \times\,\underline{13} \\ 489 \\ +\,\underline{163} \\ 2119 \end{array}$$

Thus, (43 + 120) × 13 = 2119.

Math.Content.6.NS.4

10. A: A factor of an integer is a number that divides into it evenly (i.e. without a remainder). For example, the factors of 12 are 1, 2, 3, 4, 6, and 12, and the factors of 16 are 1, 2, 4, 8, and 16. The greatest common factor (GCF) of two integers is the largest (or greatest) integer that is a factor of both numbers. For example, the common factors of 12 and 16 are 1, 2, and 4, so their GCF is 4.

One way to find the GCF of 32 and 40 is to list their factors, and then underline the common factors.

32: <u>1</u>, <u>2</u>, <u>4</u>, <u>8</u>, 16, 32

40: <u>1</u>, <u>2</u>, <u>4</u>, 5, <u>8</u>, 10, 20, 40

Since 8 is the largest number that appears in both lists, the GCF of 32 and 40 is 8.

Math.Content.6.NS.4

11. B: The least common multiple (LCM) of two integers is the smallest (or least) integer that is a multiple of both numbers. One way to find the LCM of 8 and 12 is to list their multiples, and choose the smallest number that appears in both lists.

8: 8, 16, 24, 32, 40, 48

12: 12, 24, 36, 48, 60, 72

Since 24 is the smallest number in both lists, the LCM of 8 and 12 is 24.

Math.Content.6.NS.5

12. A: Positive and negative numbers are used to describe quantities having opposite values. If you had $12 in your pocket, you could represent this amount as the positive number, 12 (or +12). Therefore, you can represent a debt (money that you would end up giving to your friend) of $12 by the negative number −12.

Math.Content.6.NS.6a

13. B: The additive inverse (or opposite) of a number, *a*, is the number, −*a*, that you can add to *a* to get a sum of 0:

$$a + (-a) = 0$$

Since 0 + 0 = 0, the additive inverse of 0 is 0.

Math.Content.6.NS.6b

14. B: To graph points on a coordinate axis, start at the origin (the point (0,0) where the two axes meet) and move left or right based on the first coordinate (the *x*-coordinate) and then up or down based on the second coordinate (the *y*-coordinate). To graph the point (−1, 4), start at the origin and move to the left (because the value is negative) one unit because the *x*-coordinate is −1. Then move up four units because the *y*-coordinate is +4. Graph the point where you end up by drawing a dot.

Math.Content.6.NS.6c

15. B: A negative number, like −2, represents the additive inverse (or opposite) of the original number, in this case, 2. To graph a negative number, start at 0 and move to the left however many units are indicated by the number. So, for this problem, graph the point two units to the left of 0.

Math.Content.6.NS.7a

16. D: You can use a number line to evaluate an inequality between two signed numbers. For instance, to evaluate choice A, graph −5 and 2 on a number line.

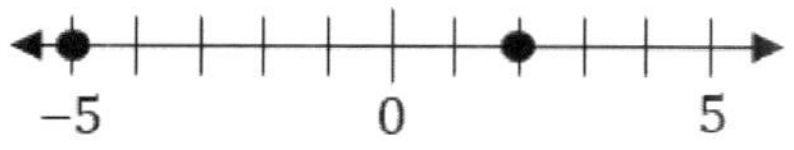

Since −5 is to the left of 2, it is less than 2, so the inequality −5 > 2 is false. Continuing in this way, you can evaluate each of the given inequalities The only one that is true is choice D, $-2 < -1$.

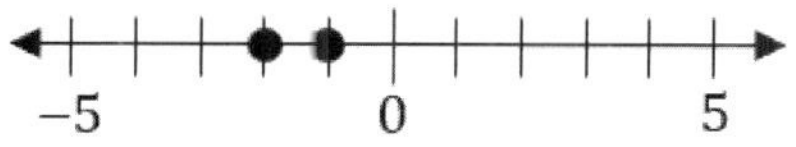

Math.Content.6.NS.7b

17. A: Positive and negative numbers are used to describe quantities having opposite values. For this problem, use negative numbers to represent distances *below* sea level and positive numbers to represent numbers *above* it. In addition, you should use a less-than sign (<) for the word "lower". Therefore, the statement "30 ft below sea level is lower than 20 ft above sea level" can be represented by the inequality $-30 < 20$.

Math.Content.6.NS.7c

18. D: The absolute value of a number (symbolized by a vertical line on each side of the number) is its distance from 0 on the number line. Since distances cannot be negative, the absolute value of a number can also never be negative. To plot the given value, –15, on a number line, graph the point 15 units to the left of 0. Therefore, the distance between –15 and 0 on a number line is 15 units, and the absolute value of –15 (notated as $|-15|$) is 15.

Math.Content.6.NS.7d

19. D: An increase in the amount of your debt means you owe more than \$50. If, for instance, your debt increases by \$10, it becomes \$60 and your account balance becomes –60. Therefore, –60 could be your new balance. (Note that a balance of –40 means that you owe \$40, which would be a decrease in your debt.)

Math.Content.6.NS.8

20. D: First graph the rectangle on a coordinate plane.

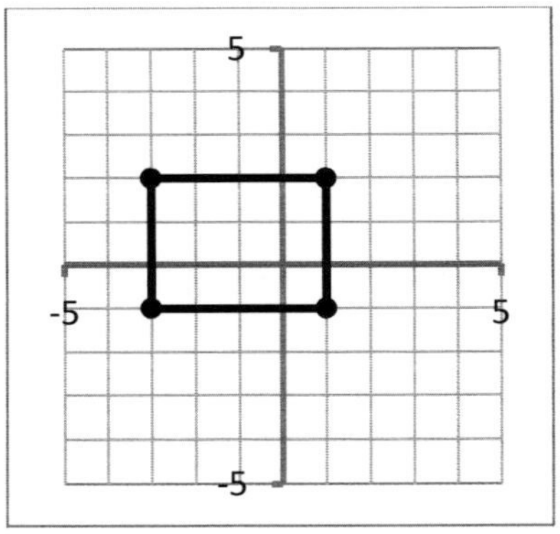

The perimeter of a rectangle is the sum of the lengths of its four sides. Notice that the height of the rectangle is 3 and its width is 4. Use this to calculate the perimeter.

$$3 + 4 + 3 + 4 = 14$$

Math.Content.6.EE.1

21. C: An exponent is a symbol that represents repeated multiplication. In the given expression, 4^3 (read "four to the third power" or "four cubed"), 4 is the base and 3 is the exponent. Using the definition just described, 4^3 can be calculated by multiplying 4 by itself three times:

$$\begin{aligned} 4^3 &= 4 \times 4 \times 4 \\ &= 64 \end{aligned}$$

Math.Content.6.EE.2a

22. A: The sum of two values is the result of adding those two values together. Therefore, the sum of x and 6 is $x + 6$.

Math.Content.6.EE.2b

23. C: According to the order of operations (PEMDAS), evaluate any quantities or expressions in parentheses first. Subtract to find the value in the parentheses and then divide 24 by the result of that subtraction:

$$\begin{aligned} 24 \div (6 - 2) &= 24 \div 4 \\ &= 6 \end{aligned}$$

Math.Content.6.EE.2c

24. C: To calculate the surface area of the cube, substitute 3 for s in the formula $A = 6s^2$:

$$A = 6 \cdot 3^2$$

Then simplify the result. According to the order of operations (PEMDAS), first calculate 3^2 and then multiply the result by 6:

$$\begin{aligned} A &= 6 \cdot 3^2 \\ &= 6 \cdot 9 \\ &= 54 \end{aligned}$$

Math.Content.6.EE.3

25. B: Since the two terms in the expression, $3a$ and $2a$, have the same variables and the same exponents on the variables (1, in this case), they are like terms. Thus, the expression can be simplified by combining like terms. To combine like terms, add the coefficients and keep the variables the same:

$$3a + 2a = 5a$$

Math.Content.6.EE.4

26. C: Two expressions are equivalent if they are always equal no matter what value you substitute for the variable(s). Thus, to determine which expression is not equivalent to $4x + 2(x + 1)$, try one or more values of x in each choice and see which one does not return the same value as $4x + 2(x + 1)$. Start with the value $x = 1$. Calculate the given expression when $x = 1$:

$$\begin{aligned} 4(1) + 2(1 + 1) &= 4 + 2(2) \\ &= 4 + 4 \\ &= 8 \end{aligned}$$

Now calculate the value of each choice when $x = 1$.

Expression	$x = 1$
$6x + 2$	$6(1) + 2 = 6 + 2$ $= 8$

Expression	$x = 1$
$2(3x + 1)$	$2(3 \cdot 1 + 1) = 2(3 + 1)$ $= 2(4)$ $= 8$
$3(2x + 1)$	$3(2 \cdot 1 + 1) = 3(2 + 1)$ $= 3(3)$ $= 9$
$2x + 2 + 4x$	$2x + 2 + 4x = 2(1) + 2$ $+ 4(1)$ $= 2 + 2 + 4$ $= 8$

Since $3(2x + 1) = 9$ when $x = 1$, the correct answer is choice C.

Alternatively, simplify the given expression:

$$4x + 2(x + 1) = 4x + 2x + 2 = 6x + 2 = 2(3x + 1)$$

(The last step is factoring 2 out from each term.) Since the given expression is equivalent to all three of A, B, and D, the only choice is C.

Math.Content.6.EE.5

27. A: A solution to an equation or inequality is a value of the variable(s) that makes the equation or inequality true. To determine which choice is a solution to the inequality $9 + 2x > 5x$, substitute each one for x in the inequality and simplify the result to see if the inequality is true. Start with $x = 2$.

$$\begin{aligned} 9 + 2x &> 5x \\ 9 + 2(2) &> 5(2) \\ 9 + 4 &> 10 \\ 13 &> 10 \end{aligned}$$

Since $13 > 10$ is true, the correct answer is $x = 2$.

Alternatively, subtract $2x$ from both sides of the inequality to get $9 > 5x - 2x$, which is equivalent to $9 > 3x$. Divide both sides by 3 to get $3 > x$, or $x < 3$. The only x to choose less than 3 is 2, so A is correct.

Math.Content.6.EE.6

28. D: Since Christine drives at 60 mph, she will travel 60 miles after one hour, 120 miles after two hours, 180 miles after three hours, and so on. In other words, you can calculate the distance she travels by multiplying the how long she drives (in hours) by 60. Since t represents the amount of time she drives (in hours), the product $60 \cdot t$, or $60t$, represents the distance she drives after t hours.

Math.Content.6.EE.7

29. B: To solve an equation of the form $x + p = q$, subtract p from both sides and then simplify the result. So, to solve the given equation, subtract 4 from both sides.

$$x + 4 = 8$$
$$x + 4 - 4 = 8 - 4$$
$$x = 4$$

Math.Content.6.EE.8

30. C: A solution to an inequality is a value of the variable(s) that makes the inequality true. An inequality of the form $x > c$ or $x < c$ will have infinitely many solutions. To graph the solutions on a number line, first graph the point c with an open circle. Since the given inequality is $x < 3$, graph the point 3 on the number line:

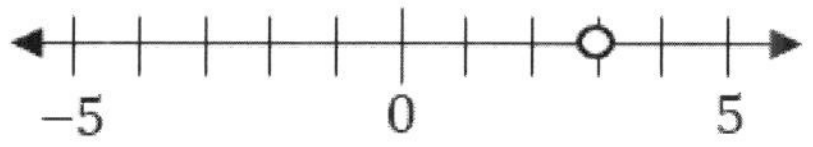

Since the solutions are every number *less than* 3, graph a line that starts at 3 and goes to the left. Put an arrow at the end of the line to show that it continues without end.

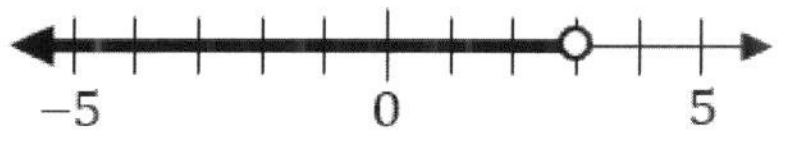

Math.Content.6.G.2

31. B: The volume of a rectangular prism is given by the formula $V = LWH$, where L, W, and H are the length, width, and height of the prism, respectively. Substitute the given dimensions into the formula and simplify the result:

$$\begin{aligned} V &= LWH \\ &= \left(2\frac{1}{2}\right)\left(1\frac{1}{3}\right)(3) \\ &= \left(\frac{5}{2}\right)\left(\frac{4}{3}\right)(3) \\ &= \frac{60}{6} \\ &= 10 \end{aligned}$$

Math.Content.6.EE.9

32. A: Since the pool is being filled at a rate of 50 gallons per minute, we can create a table showing the amount of water in the pool (in gallons) after a certain amount of time of filling by multiplying the number of minutes by 50:

Minutes	0	1	2	3	4	5
Gallons	0	50	100	150	200	250

To make the graph, plot the points (0, 0), (1, 50), (2, 100), (3, 150), and so on, until you get to (10, 500). Then connect the points with a straight line.

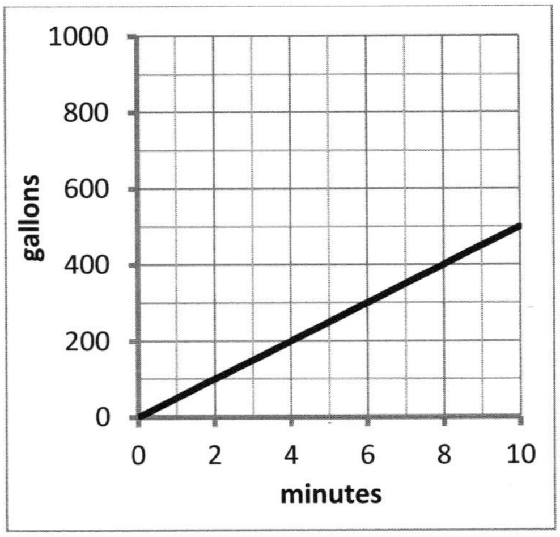

Math.Content.6.G.1

33. B: The area of a triangle is given by the formula $A = \frac{1}{2}bh$, where b is the base of the triangle and h is the height of the triangle (drawn as a line perpendicular to the base). Since the triangle is a right triangle, you can use the horizontal leg for the base and the vertical leg for the height:

$$\begin{aligned} A &= \frac{1}{2}bh \\ &= \frac{1}{2}(4)(5) \\ &= \frac{1}{2}(20) \\ &= 10 \end{aligned}$$

Math.Content.6.G.1

34. C: To calculate the area of the kite, notice that it is composed of four right triangles. As a result, you can calculate the area of each triangle and add these numbers together.

The area of a triangle is given by the formula $A = \frac{1}{2}bh$, where b is the base of the triangle and h is the height of the triangle (based on a line perpendicular to the base). Since the triangles are all right triangles, you can use the horizontal leg of each for the base and the vertical leg of each for the height.

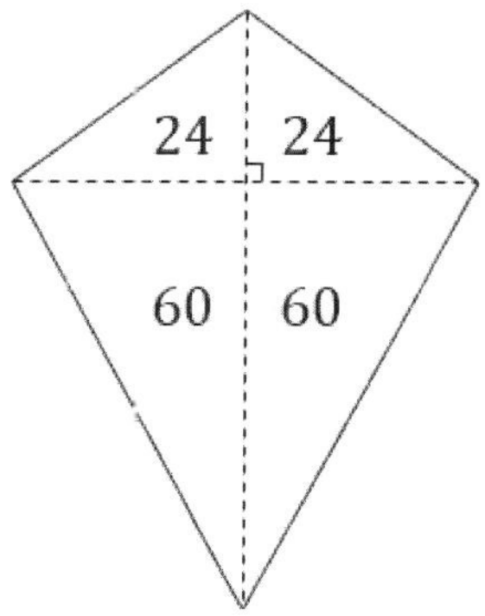

Finally, add up the four areas to get the area of the whole kite.

$$24 + 24 + 60 + 60 = 168$$

Math.Content.6.G.3

35. D: First graph the rectangle on a coordinate plane.

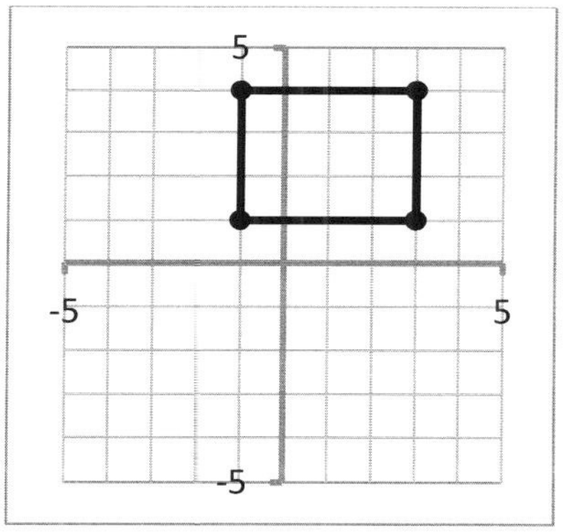

The area of a rectangle is given by the formula $A = LW$. Notice that the horizontal side, the length, of the rectangle is 4 and the vertical side, its width, is 3. Use these values to calculate its area:

$$\begin{aligned} A &= LW \\ &= (4)(3) \\ &= 12 \end{aligned}$$

Math.Content.6.G.4

36. C: A net is a two-dimensional figure that can be folded into a three-dimensional figure. For this problem, the three-dimensional figure is a square pyramid, which is a pyramid with a square base and four sides that come together at one point at the top. This figure has 1 square side, the base, and 4 triangular sides. Therefore, its net will have one square face and four triangular faces. Only choice C has this combination of faces.

Math.Content.6.SP.1

37. D: A statistical question asks a question can only be answered by gathering variable data. Choices A, B, and C are not statistical questions. To answer the question "How many students are in my English class?", you count the number of students in your class and come up with a single number. This number does not vary. Similar things can be said of "How much money is in my bank account right now?" and "How old is Tiger Woods?"

On the other hand, to answer the question about your classmates' favorite colors, you have to ask each student in your class what his favorite color is, and most likely you will get many different answers from your classmates.

Math.Content.6.SP.2

38. A: The mean of a data set is its average. Therefore, unless all of the values in the set are 12 (and you know they are not because the range is 8), some of them must be less than 12 and some must be more than 12. The range of a data set is the difference between its largest and smallest values. Since the range of this set is 8, all values in the set have to be greater than 4. If one value was 4, there has to be a value greater than 12 to ensure the average stays 12, and the difference between that value and 4 would be more than 8. Since we know the range is 8, 4 could not be a value in the set.

Math.Content.6.SP.3

39. B: The median of a data set is the middle value after the data set is put in order from least to greatest. Thus, to find the median, first you need to put the data set in order:

2, 2, 3, 4, 7, 8, 9

Since there are a total of seven values, the middle value is the fourth in the list above. Therefore, the median of the data set is 4.

Math.Content.6.SP.4

40. D: To create a box plot, you need five values: the minimum (smallest value), the lower quartile (Q1), the median (Q2), the upper quartile (Q3), and the maximum (largest value). To find these values, first put the data set in order from least to greatest:

2, 2, 3, 5, 5, 5, 6, 7, 8, 9, 10

From this, you can immediately tell that the minimum is 2 and the maximum is 10.

Next find the median (Q2) of the data. Since there are eleven values, the median is the sixth value in the set, 5. This value splits the data set into two halves (not including the median):

2, 2, 3, 5, 5 and 6, 7, 8, 9, 10

The lower quartile (Q1) is the median of the lower half, which is 3, and the upper quartile (Q3) is the median of the upper half, which is 8.

Finally, plot these five values with a box plot using a dot for the minimum (2) and maximum (10) and a box that extends from Q1 (3) to Q3 (8) with a line through it at Q2 (5):

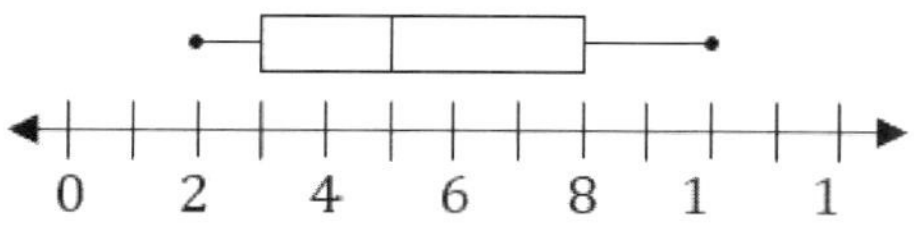

Math.Content.6.SP.5a

41. A: A histogram is a way of graphically representing the distribution of a data set. In a histogram, the data range is divided into intervals so that every value falls in one interval. For the given test scores, one natural way to set the intervals is 50 – 59, 60 – 69, 70 – 79, 80 – 89, and 90 – 100

Next calculate the number of values that fall within each interval, i.e. ask how many values are between 50 and 59 (including 50 and 59), how many are between 60 and 69, and so on:

Interval	Frequency
50 – 59	1
60 – 69	0
70 – 79	4
80 – 89	5
90 – 100	2

Finally, display this information graphically using a vertical bar for each interval and making its height equal to the frequency of the interval:

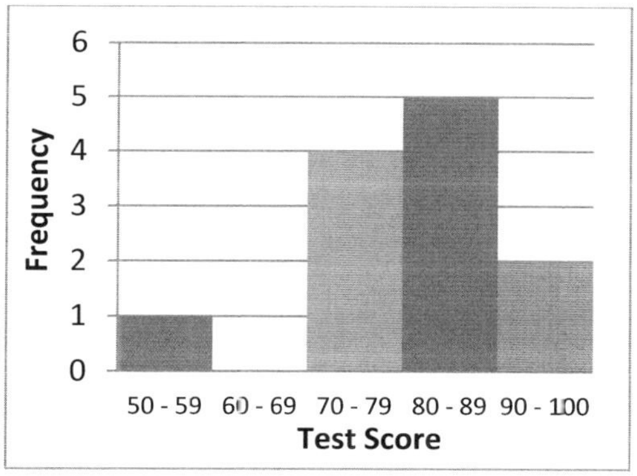

Math.Content.6.SP.5b

42. C: The appropriate unit of measure of a quantity corresponds to what you are counting when you measure it. For instance, if the price of a car is $40,000, because you are counting dollars, the unit of measure of the quantity $40,000 is dollars. To measure the attendance of the Broadway shows, you count the number of people who go to each show. Therefore, the unit of measure is people.

Math.Content.6.SP.5c

43. A: The interquartile range of a data set is the difference between the upper and lower quartiles (Q3 and Q1). To calculate it, first put the data set in order from least to greatest:

11, 14, 15, 15, 16, 19, 27

Since there are seven values, the median is the fourth value in the set, 15. This value splits the data set into two halves (not including the median):

11, 14, 15 and 16, 19, 27

The lower quartile is the median of the lower half, which is 14, and the upper quartile is the median of the upper half, which is 19. Subtract these values to find the interquartile range:

$$19 - 14 = 5$$

Math.Content.6.SP.5d

44. B: Data is skewed if values on one side tend to be farther from the 'center' than values on the other side. The graphs below show data skewed to the left and right, respectively.

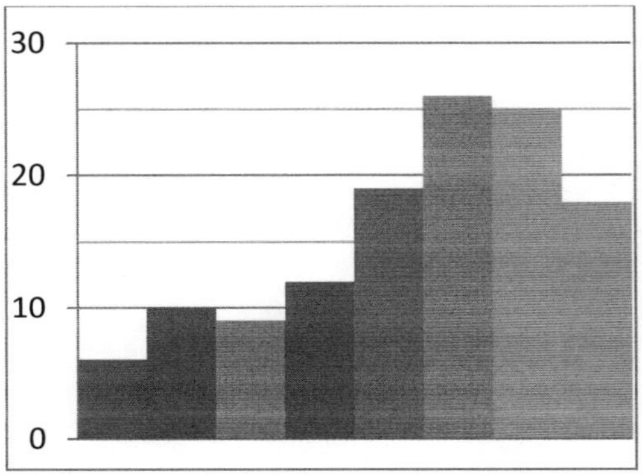

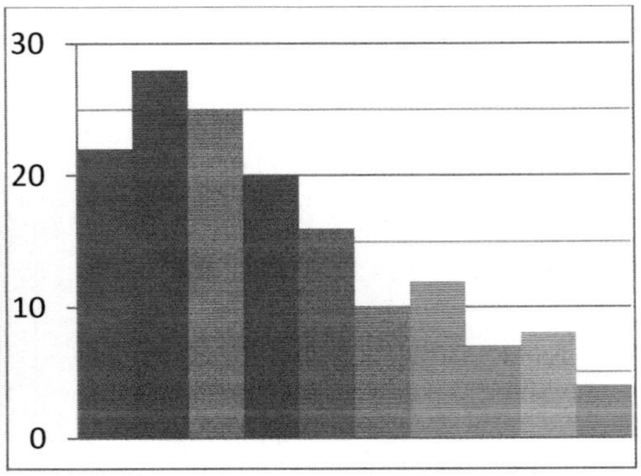

In these situations, the mean get 'pulled away' from what we think of as the more natural value of the center of the data. In the graphs above, the mean gets pulled to the left in the left graph and pulled to the right in the right one. Consequently, in skewed data, the median is usually thought of as a better representative of the center of the data than the mean. Mode, while a measure of the 'center', is most helpful when the data values are not numeric. Since they are in this case (employee salaries), the median is still the most useful. (Range is not a measure of center, but a measure of spread.)

Mathematics Practice Test #2

1. A classroom has three girls for every boy. Write the ratio of the number of girls to the number of boys.

a. 1:1
b. 1:3
c. $3:\frac{1}{3}$
d. 3:1

2. A pancake recipe calls for 2 cups of all-purpose flour to make 20 pancakes. How many cups of flour are used to make each pancake?

a. $\frac{1}{10}$ cups
b. 9 cups
c. 10 cups
d. 18 cups

3. A man can read 2 pages every 3 minutes. The table below shows the number of pages he can read after four different periods of time.

Pages	Minutes
2	3
4	6
6	9
8	

What is the missing value in the table?

a. 10
b. 12
c. 15
d. 18

4. A certain car factory can produce 9 cars every 4 minutes on an assembly line. At this rate, how many cars will it produce in 20 minutes?

a. 36 cars
b. 40 cars
c. 45 cars
d. 48 cars

5. What is 20% of 60?

a. 12
b. 15
c. 18
d. 20

6. An object that weighs one pound has a mass of about 0.454 kg. What is the weight, in pounds, of an object with a mass of 3 kg?

a. 0.1513 lb
b. 1.362 lb
c. 2.546 lb
d. 6.608 lb

7. A recipe calls for 1/3 of a cup of cocoa powder to make a pan of brownies. How many pans can you make with 1/4 of a cup of cocoa powder?

a. 1/2 of a pan
b. 3/4 of a pan
c. 4/3 of a pan
d. 2 pans

8. Calculate $825 \div 33$.

a. 24
b. 25
c. 34
d. 35

9. Calculate $26 \times 105 - 40$.

a. 1690
b. 2560
c. 2690
d. 2770

10. What is the greatest common factor (GCF) of 30 and 45?

a. 3
b. 5
c. 15
d. 90

11. What is the least common multiple (LCM) of 6 and 12?

a. 12
b. 24
c. 30
d. 36

12. The Dead Sea is 423 m below sea level. Represent this elevation as a signed number.

a. -423
b. 0
c. $\frac{1}{423}$
d. 423

13. Simplify $-(-6)$.

a. -6
b. -3
c. 3
d. 6

14. Graph the point (3, –2) on a coordinate plane.

a. 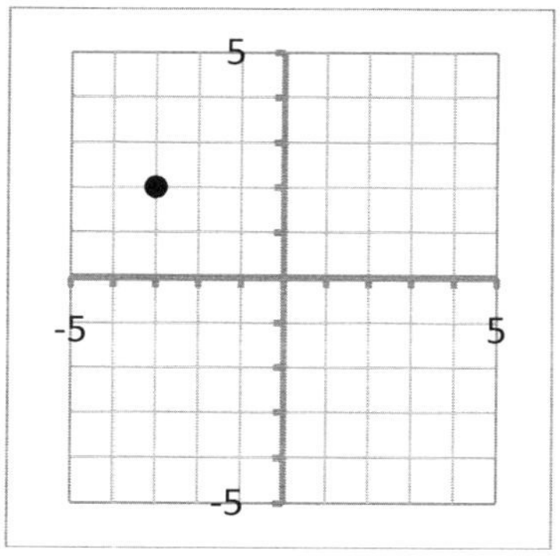

b. 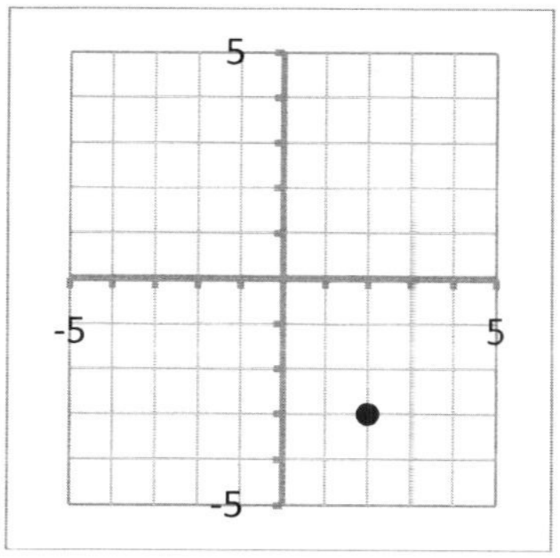

c. 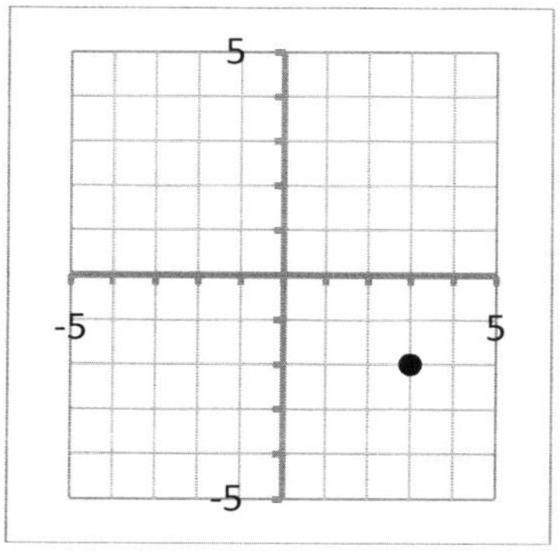

d. 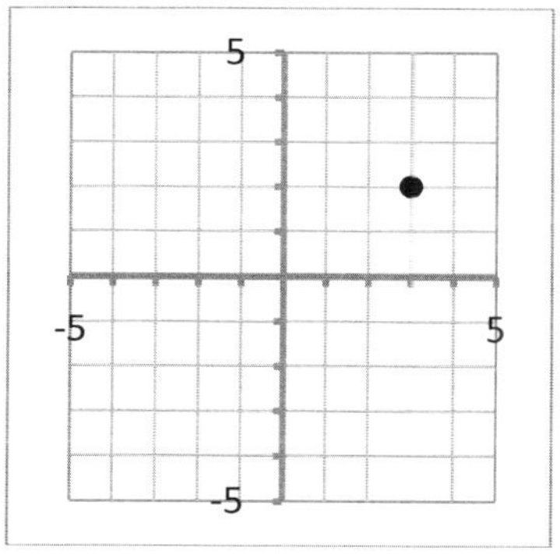

15. Graph the number –3 on a number line.

a. 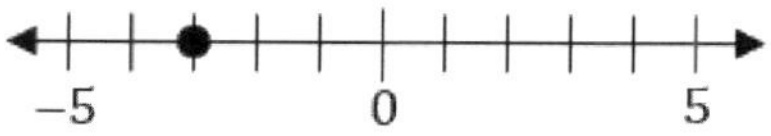

b. 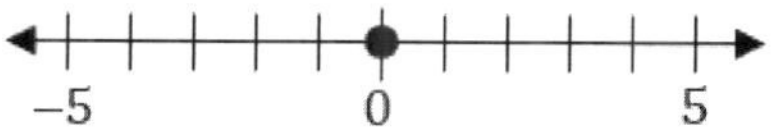

c. 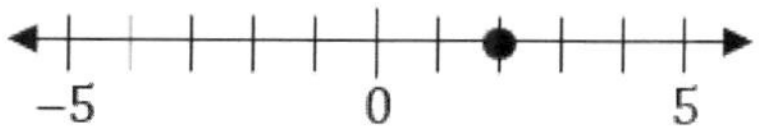

d. 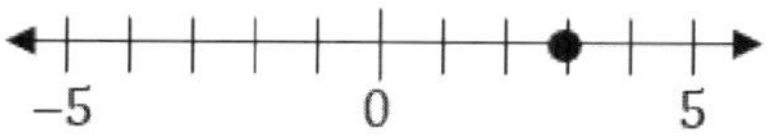

16. Which of the following is a true inequality?

a. $-5 > 0$
b. $-3 < 2$
c. $-2 > -1$
d. $4 < -3$

17. What does the inequality 15°C > –5°C represent?

a. 15°C is colder than 5°C.
b. 15°C is warmer than 5°C.
c. 15°C is colder than –5°C.
d. 15°C is warmer than –5°C.

18. Calculate |8|.

a. –8
b. 0
c. 4
d. 8

19. The balance on your savings account is –80, which means that have $80 of debt. You decrease the amount of your debt by adding more money into it. If you still have debt after you've added this money, which of the following could be your new balance?

a. 90
b. 70
c. –70
d. –90

20. The vertices of a rectangle are (–1, –2), (–1, 4), (2, 4), and (2, –2). What is its perimeter?

a. 3
b. 6
c. 9
d. 18

21. Evaluate 2^4.

a. 8
b. 16
c. 24
d. 32

22. Write a mathematical expression for the statement, "Subtract *y* from 7."

a. $-7y$
b. $7 - y$
c. $y + 7$
d. $y - 7$

23. Calculate $12 - (2 \times 4)$.

a. 4
b. 10
c. 24
d. 40

24. Calculate the value of $2x^3$ when $x = 3$.

a. 18
b. 54
c. 144
d. 216

25. Distribute $4(2 + 3y)$.

a. $12y$
b. $20y$
c. $8 + 3y$
d. $8 + 12y$

26. Which expression is NOT equivalent to $4n + 6$?

a. $2(2n + 3)$
b. $2(n + 3) + n$
c. $n + 3(n + 2)$
d. $3n + 6 + n$

27. Which of the following is a solution to the equation $(x + 2)^2 = 9x$?

a. 2
b. 3
c. 4
d. 5

28. Ricky is three years younger than Craig. If Craig is *n* years old, how old is Ricky?

a. $n - 3$
b. $n + 3$
c. $3n$
d. $\frac{n}{3}$

29. Solve the equation $5x = 10$.

a. 2
b. 5
c. 15
d. 50

30. Graph the solutions to the inequality $x > -1$ on a number line.

a. 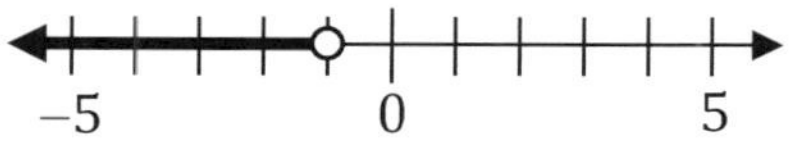

b. 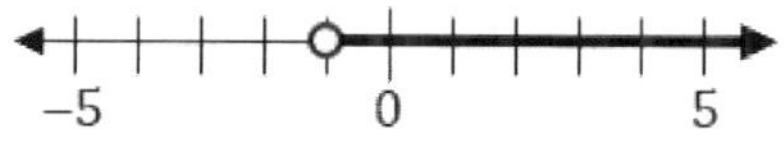

c. 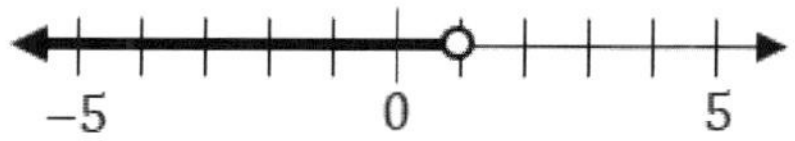

d. 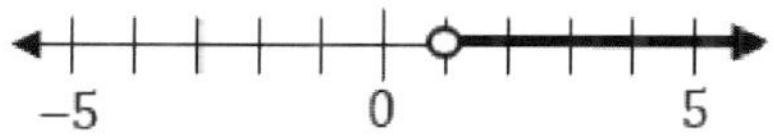

31. Calculate the area of the triangle below.

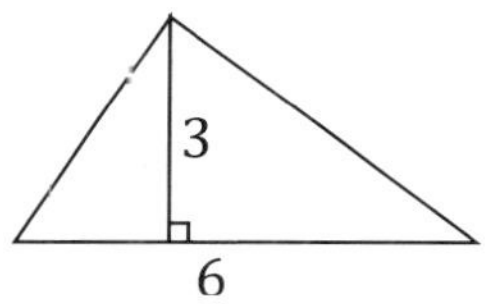

a. 3
b. 9
c. 12
d. 18

32. A trucker is driving at 60 mph down a highway. Make a graph showing the distance he travels in the first 8 hours.

a. 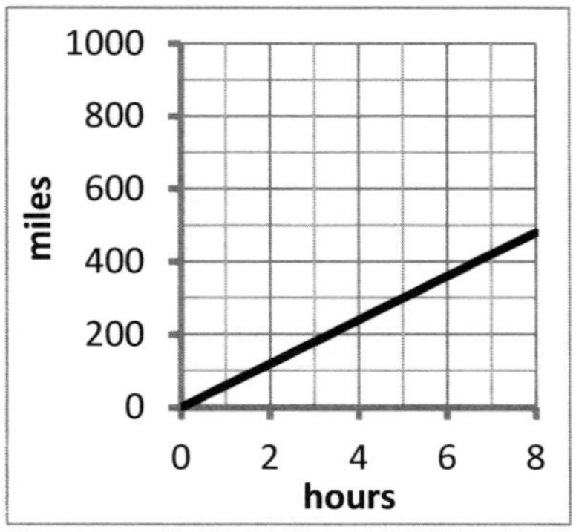

b. 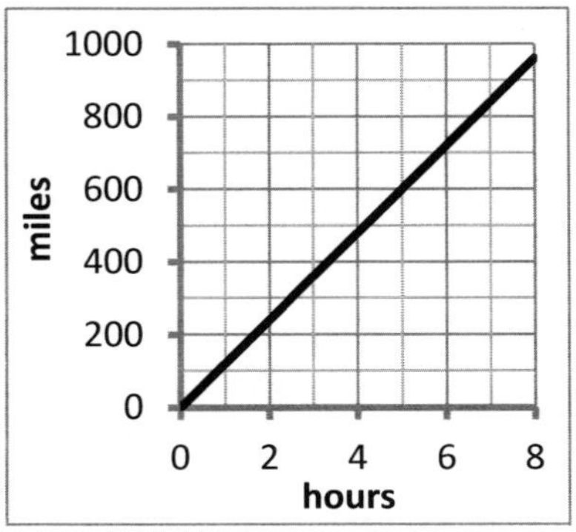

c. 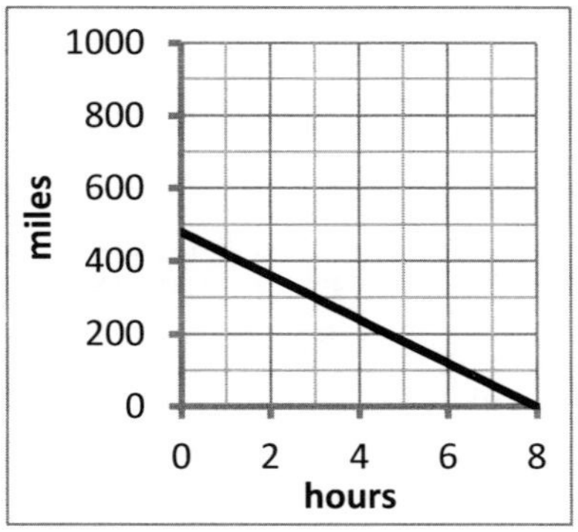

d. 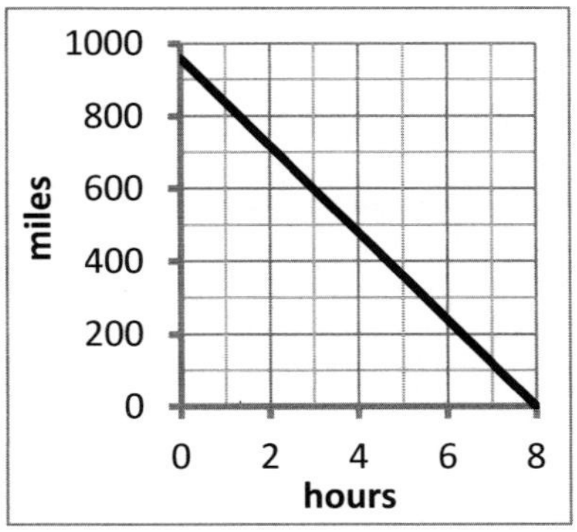

33. Calculate the area of the right trapezoid below.

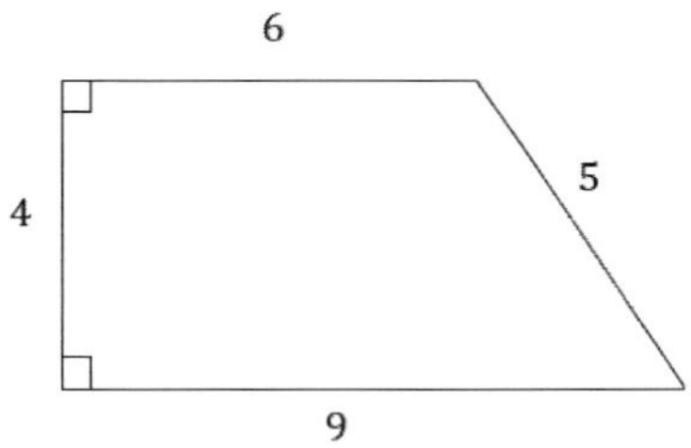

a. 30
b. 36
c. 37.5
d. 45

34. A rectangular room is $14\frac{1}{2}$ feet long, 20 feet wide, and $12\frac{1}{2}$ feet tall. What is its volume?

a. 2700 cubic feet
b. 2850 cubic feet
c. 3625 cubic feet
d. 3875 cubic feet

35. The vertices of a rectangle are (0, 2), (3, 2), (3, –4), and (0, –4). What is its area?

a. 9
b. 12
c. 15
d. 18

36. Which of the following is a net of a cube?

a. 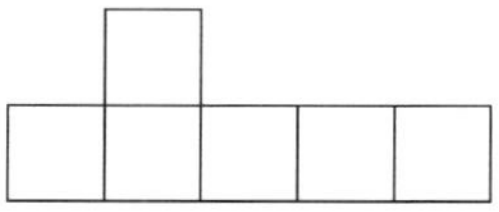

b. 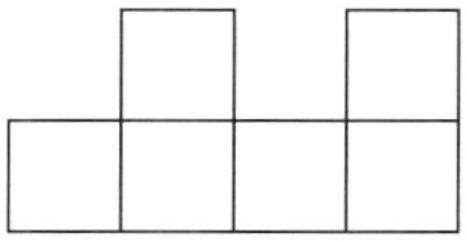

c. 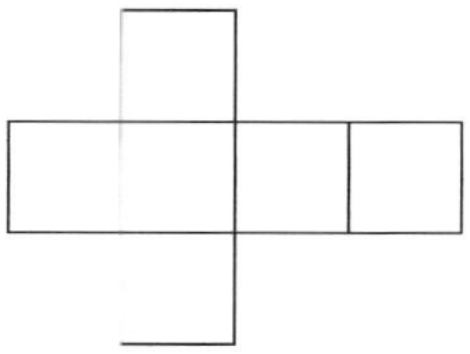

d. 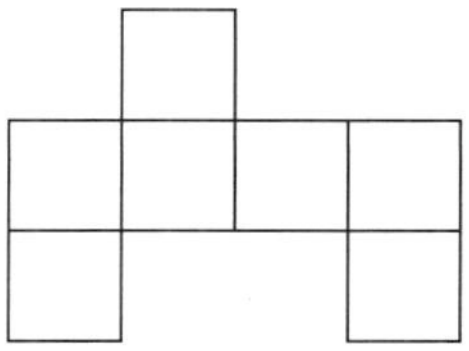

37. Which of the following is a statistical question?

a. How much money did my father make last year?
b. How many pets do different families in my town have?
c. How tall is Shaquille O'Neal?
d. Who won the baseball game last night?

38. The median of a data set is 9, and its range is 6. Which of the following values could NOT be value in the set?

a. 6
b. 9
c. 12
d. 16

39. Calculate the mean of the data set: 2, 9, 1, 9, 9, 2, 3.

a. 3
b. 5
c. 6
d. 9

40. Create a box plot for the data set: 12, 20, 19, 16, 17, 20, 20, 21, 18, 18, 17.

a. 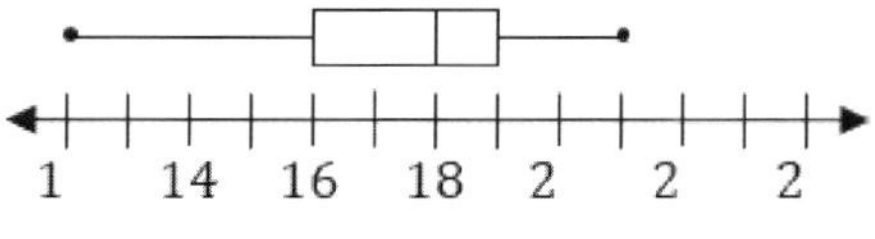

b. 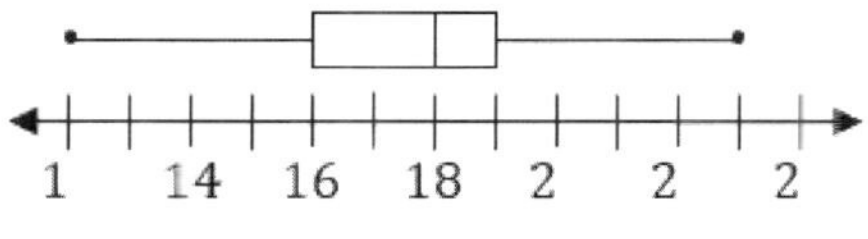

c. 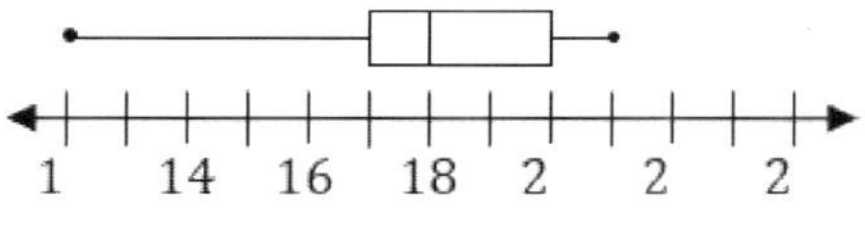

d. 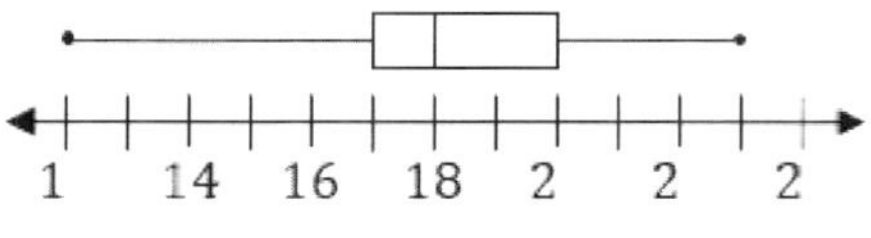

41. The 200-meter dash times (in seconds) for runners at Engelbert High School are shown below.

25.6, 27.1, 23.1, 23.8, 24.8, 26.8, 26.3, 24.1, 24.8, 24.0

Create a histogram to display the data.

a. b. c. d.

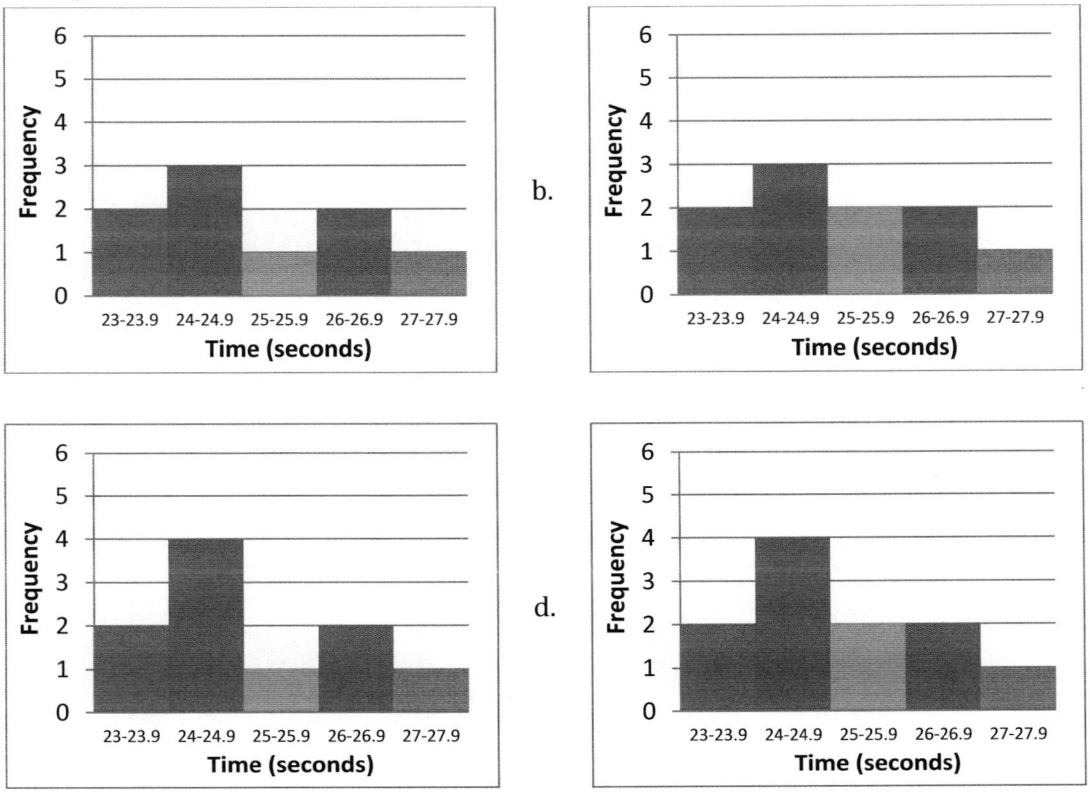

42. What unit of measure would be appropriate to use when investigating the salary of CEOs?

a. companies
b. dollars
c. people
d. years

43. Calculate the interquartile range of the data set: 2, 0, 5, 12, 9, 12, 4.

a. 5
b. 6
c. 10
d. 12

44. You notice that the distribution of the test scores of your classmates is skewed to the left. Which of the following would be the best choice for the measure of center of the data?

a. mean
b. median
c. mode
d. range

Answer Key and Explanations for Test #2

Math.Content.6.RP.1

1. D: A ratio is a relationship between two numbers of the same type. For instance, if the ratio of apples to oranges in a basket is 3:2, then the basket contains 3 apples for every 2 oranges. In the given problem, we are looking for ratio between the numbers of girls and boys. Since we are told that there are three girls for every one boy, the correct ratio is 3:1.

Math.Content.6.RP.2

2. A: In the recipe, the ratio of the number of cups of flour to the number of pancakes is 2:20. To find the unit rate (in this case, the amount of flour per pancake) write this ratio as a fraction. The result is $\frac{2}{20}$, which can be simplified to $\frac{1}{10}$. Therefore, $\frac{1}{10}$ cup of flour is used to make each pancake.

Math.Content.6.RP.3a

3. B: You should notice that the four numbers in the left column (2, 4, 6, 8) increase by 2 from top-to-bottom. At the same time, the numbers in the right column (3, 6, 9) increase by 3 from top-to-bottom. Since 9 + 3 = 12, the missing value is 12.

Math.Content.6.RP.3b

4. C: To find the average rate of the factory, divide the number of cars it can produce by the time it takes:

$$\frac{9\text{ cars}}{4\text{ minutes}} = 2.25\text{ cars per minute}$$

Therefore, the average rate is 2.25 cars per minute. To determine how many cars it will produce in 20 minutes, multiply 2.25 by 20:

$$2.25\text{ cars per minute} \times 20\text{ minutes} = 45\text{ cars}$$

Math.Content.6.RP.3c

5. A: A percent represents a part out of 100. For example, 20% is equivalent to 20 out of 100, or $\frac{20}{100}$. To find 20% of 60, multiply $\frac{20}{100}$ by 60:

$$\frac{20}{100} \times 60 = 12$$

Math.Content.6.RP.3d

6. D: We are told that the ratio of pounds to kilograms is 1:0.454. To convert kilograms to pounds, write this ratio as a fraction with pounds in the numerator (i.e. on top), and then multiply that fraction by the mass in kilograms:

$$\frac{1\text{lb}}{0.454\text{kg}} \times 3\text{kg} = \frac{3}{0.454}\text{lb}$$
$$\approx 6.608lb$$

Math.Content.6.NS.1

7. B: Set up a proportion: if one pan calls for 1/3 of a cup, how many pans call for ¼ of a cup?

$$\frac{1\,pan}{{}^{1}/_{3}\;cup} = \frac{x\;pans}{{}^{1}/_{4}\;cup} \xrightarrow{yields} \frac{1}{4} = \frac{1}{3}x$$

Therefore, to determine how many pans you can make, divide 1/4 by 1/3:

$$\frac{1}{4} \div \frac{1}{3}$$

To divide a number by a fraction, (1) change the division sign to a multiplication sign and (2) invert the (second) fraction by exchanging its numerator and denominator. Then perform the multiplication normally:

$$\begin{aligned}\frac{1}{4} \div \frac{1}{3} &= \frac{1}{4} \times \frac{3}{1} \\ &= \frac{3}{4}\end{aligned}$$

Therefore, you can make 3/4 of a pan.

Math.Content.6.NS.2

8. B: Perform long division to divide the dividend (in this case, 825) by the divisor (33) using a division sign called a tableau:

$$33 \overline{)825}$$

Start by trying to divide the first digit in the dividend, 8, by 33. Since it is too small, add the next digit. Dividing 82 by 33 gives you 2 with a remainder. Put the 2 at the top of the tableau above the 2, and then multiply 33 by 2 and subtract to find the remainder:

```
      2
33 | 825
-    66
     16
```

Next, bring down the next digit in 825 (which is a 5), and repeat the last step by dividing 165 by 33:

```
      25
33 | 825
-    66↓
     165
-    165
       0
```

Therefore, 825 ÷ 33 = 25.

Math.Content.6.NS.3

9. C: By order of operations (PEMDAS), perform any multiplication or division before any addition or subtraction. So, perform the multiplication first:

$$\begin{array}{r} 105 \\ \times \underline{\ \ 26} \\ 630 \\ + \underline{2100} \\ 2730 \end{array}$$

Therefore, $26 \times 105 - 40 = 2730 - 40$. Perform the subtraction:

$$\begin{array}{r} 2730 \\ - \underline{\ \ \ 40} \\ 2690 \end{array}$$

Thus, $26 \times 105 - 40 = 2690$.

Math.Content.6.NS.4

10. C: A factor of an integer is a number that divides into it evenly (i.e. without a remainder). For example, the factors of 12 are 1, 2, 3, 4, 6, and 12, and the factors of 16 are 1, 2, 4, 8, and 16. The greatest common factor (GCF) of two integers is the largest (or greatest) integer that is a factor of both numbers. For example, the common factors of 12 and 16 are 1, 2, and 4, so their GCF is 4.

One way to find the GCF of 30 and 45 is to list their factors, and then underline the common factors:

30: <u>1</u>, 2, <u>3</u>, <u>5</u>, 6, 10, <u>15</u>, 30

45: <u>1</u>, <u>3</u>, <u>5</u>, 9, <u>15</u>, 45

Since 15 is the largest number that appears in both lists, the GCF of 30 and 45 is 15.

Math.Content.6.NS.4

11. A: The least common multiple (LCM) of two integers is the smallest (or least) integer that is a multiple of both numbers. One way to find the LCM of 6 and 12 is to list their multiples, and choose the smallest number that appears in both lists:

6: 6, 12, 18, 24, 30, 36

12: 12, 24, 36, 48, 60, 72

Since 12 is the smallest number in both lists, the LCM of 6 and 12 is 12.

Math.Content.6.NS.5

12. A: Positive and negative numbers are used to describe quantities having opposite values. If you are 200 m *above* sea level, you could represent this elevation as the positive number, 200 (or +200). Similarly, you can represent 423 m *below* sea level by the negative number, –423.

Math.Content.6.NS.6a

13. D: A negative number, like –10, represents the additive inverse (or opposite) of the original number, in this case, 10. Therefore, the given expression, –(–10) is opposite of –6. In other words, it's the opposite of the opposite of 6, so it's equal to 6.

Math.Content.6.NS.6b

14. C: To graph points on a coordinate axis, start at the origin, the point (0,0) where the two axes meet, and move left (if it's negative) or right (if it's positive) based on the first coordinate (the *x*-coordinate) and then up or down (if positive or negative, respectively) based on the second coordinate (the *y*-coordinate). To graph the point (3, –2), start at the origin and move right three units because the *x*-coordinate is +3. Then move down two units because the *y*-coordinate is –2. Graph the point where you end up by drawing a dot.

Math.Content.6.NS.6c

15. A: A negative number, like –3, represents the additive inverse (or opposite) of the original number, in this case, 3. To graph a negative number, start at 0 and move to the left however many units are indicated by the number. So, for this problem, graph the point three units to the left of 0.

Math.Content.6.NS.7a

16. B: You can use a number line to evaluate an inequality between two signed numbers. For instance, to evaluate choice A, graph –5 and 0 on a number line.

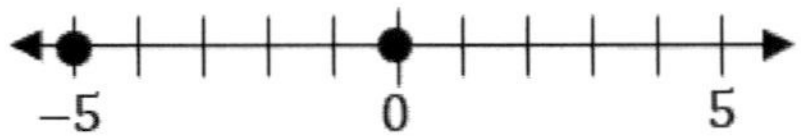

Since –5 is to the left of 0, it is less than 0, so the inequality –5 > 0 is false. Continuing in this way, you can evaluate each of the given inequalities. The only one that is true is choice B, –3 < 2, since -3 is to the left of 2.

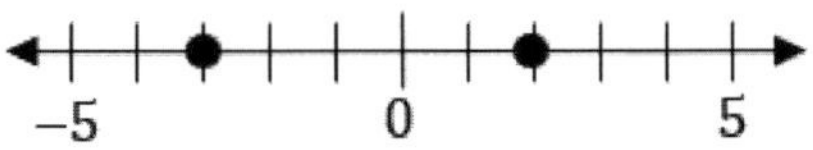

Math.Content.6.NS.7b

17. D: Positive and negative numbers are used to describe quantities having opposite values. In Celsius, positive values represent degrees *above* the freezing point of water and negative values represent degrees below it. Moreover, if one value is greater than another, then it corresponds to being warmer than the other. Therefore, the inequality 15°C > –5°C represents the statement "15°C is warmer than –5°C."

Math.Content.6.NS.7c

18. D: The absolute value of a number (symbolized by a vertical line on each side of the number) is its distance from (how far away it is from) 0 on the number line. Since distances cannot be negative, the absolute value of a number also cannot be negative. To plot the given value, 8, on a number line, graph the point 8 units to the right of 0 (to the right because 8 is positive). Then, the distance between 8 and 0 on a number line is 8 units, and the absolute value of 8 (notated as $|8|$) is 8.

Math.Content.6.NS.7d

19. C: A decrease of the amount of your debt means you owe less than \$80. If, for instance, you decrease your debt by \$10, it becomes a debt of \$70 and your account balance becomes −70. Therefore, −70 could be your new balance. (Note that a balance of −90 means that you owe \$90, which would be an increase in your debt. Also note that although you could add \$150 to your account – so that you have an extra \$70 instead of owing money – we are told that you still have debt after you've added money, so the answer here has to be a negative number.)

Math.Content.6.NS.8

20. D: First graph the rectangle on a coordinate plane.

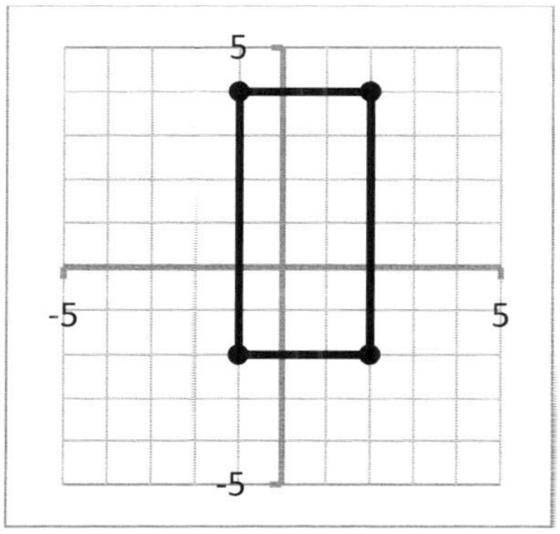

The perimeter of a rectangle is the sum of the lengths of its four sides. Notice that the height of the rectangle is 6 and its width is 3. Use this to calculate the perimeter:

$$6 + 3 + 6 + 3 = 18$$

Math.Content.6.EE.1

21. B: An exponent is a symbol that represents multiplying the base by itself a certain number of times. In the given expression, 2^4 (read "two to the fourth power"), 2 is the base and 4 is the exponent. This means that 2^4 can be calculated by multiplying 2 by itself four times:

$$\begin{aligned} 2^4 &= 2 \times 2 \times 2 \times 2 \\ &= 16 \end{aligned}$$

Math.Content.6.EE.2a

22. B: When you subtract one number *from* another, the second number listed comes first, then a subtraction sign (−), and finally the first number in the statement. Therefore, "Subtract y from 7" can be translated into $7 - y$.

Math.Content.6.EE.2b

23. A: According to the order of operations (PEMDAS), first multiply to find the value in the parentheses and then subtract the result from 12:

$$\begin{aligned} 12 - (2 \times 4) &= 12 - 8 \\ &= 4 \end{aligned}$$

Math.Content.6.EE.2c

24. B: First substitute 3 for x in the given expression $2x^3$:

$$2 \cdot 3^3$$

Then simplify the result. According to the order of operations (PEMDAS), first calculate 3^3 and then multiply the result by 2:

$$\begin{aligned} 2 \cdot 3^3 &= 2 \cdot 27 \\ &= 54 \end{aligned}$$

Math.Content.6.EE.3

25. D: The distributive property, $a(b + c) = ab + ac$, is a rule for simplifying the product of a single term and a sum. In this case, you can distribute 4 over the sum in parentheses. To do this, multiply each term inside the parentheses by 4 and simplify each product:

$$\begin{aligned} 4(2 + 3y) &= 4 \cdot 2 + 4 \cdot 3y \\ &= 8 + 12y \end{aligned}$$

Math.Content.6.EE.4

26. B: Two expressions are equivalent if they are always equal no matter what value you substitute for the variable(s). Thus, to determine which expression is not equivalent to $4n + 6$, try one or more values of x in each choice and see which one does not return the same value as $4n + 6$. Start with $x = 1$. Calculate the value of the given expression when $x = 1$:

$$\begin{aligned} 4(1) + 6 &= 4 + 6 \\ &= 10 \end{aligned}$$

Now calculate the value of each choice when $x = 1$:

Expression	$x = 1$
$2(2n + 3)$	$\begin{aligned} 2(2 \cdot 1 + 3) &= 2(2 + 3) \\ &= 2(5) \\ &= 10 \end{aligned}$
$2(n + 3) + n$	$\begin{aligned} 2(1 + 3) + n &= 2(4) \\ &= 8 \end{aligned}$
$n + 3(n + 2)$	$\begin{aligned} 1 + 3(1 + 2) &= 1 + 3(3) \\ &= 1 + 9 \\ &= 10 \end{aligned}$

Expression	$x = 1$
$3n + 6 + n$	$3(1) + 6 + 1 = 3 + 6 + 1$ $= 10$

Since $2(n + 3) + n = 8$ when $x = 1$, it is not equal to the given expression, and the correct answer is choice B.

Math.Content.6.EE.5

27. C: The solution to an equation or inequality is the value(s) of the variable(s) that makes the equation or inequality true. To determine which choice is a solution to the equation $(x + 2)^2 = 9x$, substitute each one for x in the equation and simplify the result to see if the resulting equation is true. Start with $x = 2$:

$$\begin{aligned}(x + 2)^2 &= 9x \\ (2 - 2)^2 &= 9(2) \\ (4)^2 &= 18 \\ 16 &= 18\end{aligned}$$

Since $16 \neq 18$, $x = 2$ is not a solution to the given equation. If you substitute $x = 3$ in the given equation, the result is again false, so try $x = 4$:

$$\begin{aligned}(x + 2)^2 &= 9x \\ (4 + 2)^2 &= 9(4) \\ (6)^2 &= 36 \\ 36 &= 36\end{aligned}$$

Since $36 = 36$ is true, the correct answer is $x = 4$.

Math.Content.6.EE.6

28. A: Since Ricky is three years younger than Craig, he would be 10 years old if Craig is 13, 11 years old if Craig is 14, 12 years old if Craig is 15, and so on. In other words, you can calculate Ricky's age by subtracting 3 from Craig's age. Since Craig is n years old, Ricky is $n - 3$ years old.

Math.Content.6.EE.7

29. A: To solve an equation of the form $px = q$, divide from both sides by p and then simplify the result. So, to solve the given equation, divide both sides by p:

$$\begin{aligned}5x &= 10 \\ \frac{5x}{5} &= \frac{10}{5} \\ x &= 2\end{aligned}$$

Math.Content.6.EE.8

30. B: The solution to an inequality is the value(s) of the variable(s) that makes the inequality true. An inequality of the form $x > c$ or $x < c$ will have infinitely many solutions. To graph the solutions on a number line, first graph point c with an open circle. Since the given inequality is $x > -1$, graph the point -1 on the number line:

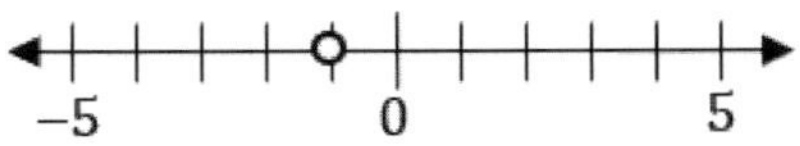

Since the solutions are every number *greater than* -1, graph a line that starts at -1 and goes to the right. Put an arrow at the end of the line to show that it continues without end.

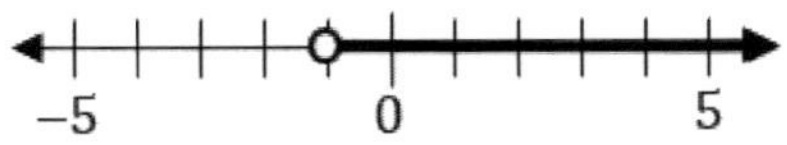

Math.Content.6.G.1

31. B: The area of a triangle is given by the formula $A = \frac{1}{2}bh$, where b is the base of the triangle and h is the height of the triangle (drawn as a line perpendicular to the base). In the given triangle, use the horizontal line at the bottom as the base and the vertical line in the middle for the height:

$$\begin{aligned} A &= \frac{1}{2}bh \\ &= \frac{1}{2}(6)(3) \\ &= \frac{1}{2}(18) \\ &= 9 \end{aligned}$$

Math.Content.6.EE.9

32. A: Since the trucker is driving at 60 mph, we can create a table showing the distance he has traveled (in miles) after a certain amount of time (in hours) of driving by multiplying the number of hours by 60:

Hours	0	1	2	3	4	5
Miles	0	60	120	180	240	300

To make the graph, plot the points (0, 0), (1, 60), (2, 120), (3, 180), and so on, until you get to (8, 480). Then connect the points with a straight line.

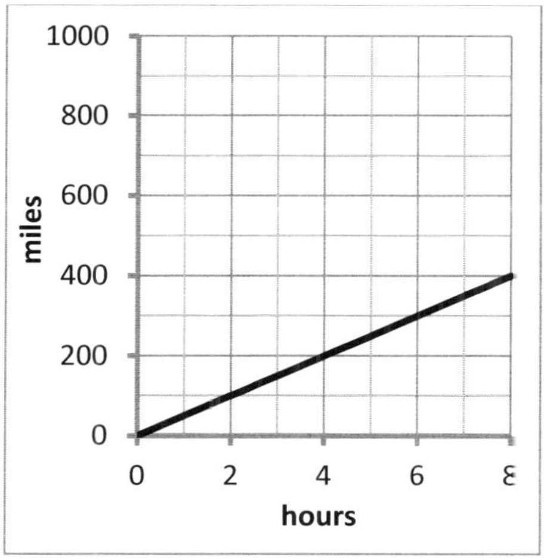

Math.Content.6.G.1

33. A: To calculate the area of the trapezoid, decompose it into two shapes, a rectangle and a right triangle, by drawing a vertical line.

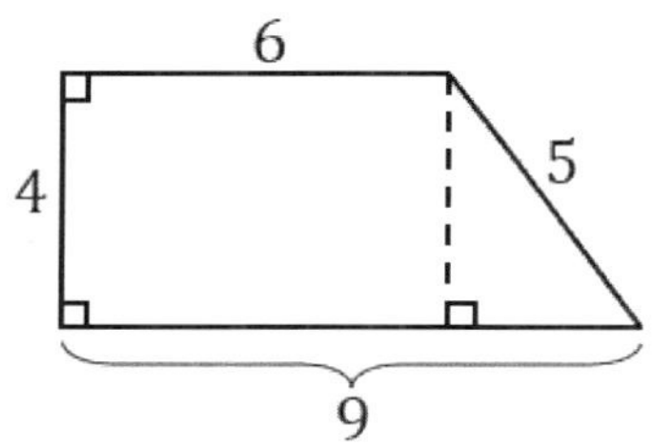

Since the left part is a rectangle, its right side is 4 units long and its bottom side is 6 units long. Moreover, this means that the base of the triangle must be 3 since the base of the original polygon was 9 and 3 + 6 = 9.

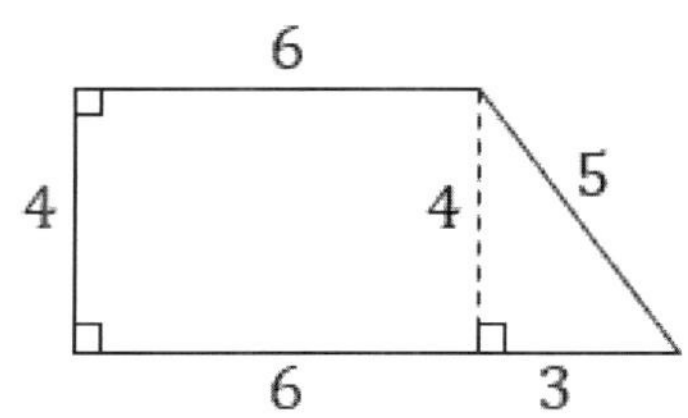

Next use the area formulas, $A = LW$ and $A = \frac{1}{2}bh$, to calculate the areas of the rectangle and triangle, respectively.

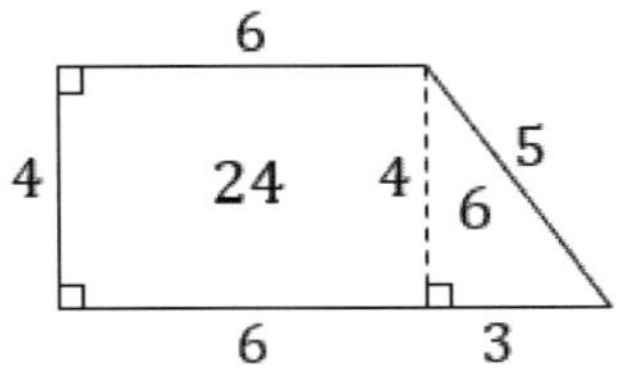

Finally, add up the two areas to get the area of the trapezoid:

$$24 + 6 = 30$$

Math.Content.6.G.2

34. C: The volume of a rectangular prism is given by the formula $V = LWH$, where L, W, and H are the length, width, and height of the prism, respectively. Substitute the given dimensions into the formula and simplify the result:

$$\begin{aligned} V &= LWH \\ &= \left(14\frac{1}{2}\right)(20)\left(12\frac{1}{2}\right) \\ &= \left(\frac{29}{2}\right)(20)\left(\frac{25}{2}\right) \\ &= \frac{14{,}500}{4} \\ &= 3625 \end{aligned}$$

Math.Content.6.G.3

35. D: First graph the rectangle on a coordinate plane.

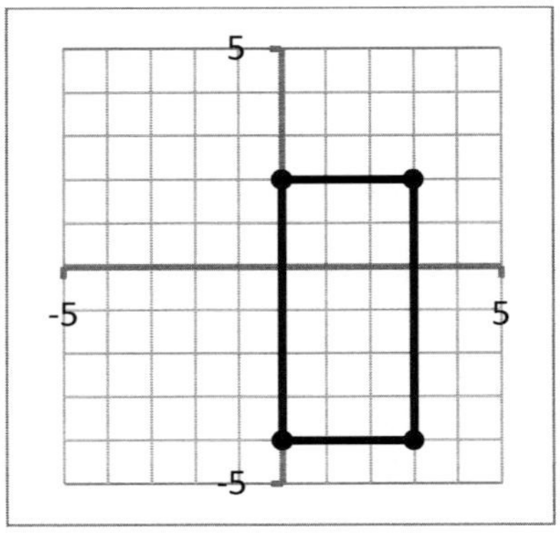

The area of a rectangle is given by the formula $A = LW$. Notice that the horizontal side (length) of the rectangle is 3 and the vertical side (width) is 6. Use these values to calculate its area:

$$\begin{aligned} A &= LW \\ &= (3)(6) \\ &= 18 \end{aligned}$$

Math.Content.6.G.4

36. C: A net is a two-dimensional figure that can be folded into a three-dimensional figure. For this problem, the three-dimensional figure is a cube. Since a cube has six square faces, its net will have six square faces (so that eliminates choice D).

In choices A and D, there are problems with the nets that make it impossible to get a cube with a bottom and a top face. Only in choice C can you fold the four horizontal squares in the middle to get the sides of the cube and then fold the remaining two squares to get the top and bottom of the cube.

Math.Content.6.SP.1

37. B: A statistical question asks a question that can only be answered by gathering variable data. Choices A, C, and D are not statistical questions. To answer the question, "How much money did my father make last year?" you ask your father how much he made, and he gives you a single number. This number does not vary. Similar things can be said of "How tall is Shaquille O'Neal?" and "Who won the baseball game last night?"

On the other hand, to answer the question "How many pets do different families in my town have?", you have to ask a member of each family (or at least a sample) how many pets they have, and most likely you will get many different answers from the families.

Math.Content.6.SP.2

38. D: The median of a data set is the middle value after the data set is put in order from least to greatest. Therefore, we can expect at least one of the values to be 9.

The range of a data set is the difference between its largest and smallest values. Since the range of this set is 6, no value in the set can be more than 15. If it was, then the difference between 9 and that number would be more than 6.

Math.Content.6.SP.3

39. B: The mean of a data set is its average and can be calculated by adding the values in the set and dividing the sum by the number of data points. First add up the data values:

$$2 + 9 + 1 + 9 + 9 + 2 + 3 = 35$$

Then divide the sum by 7 since there are seven data values:

$$Mean = \frac{35}{7} = 5$$

Math.Content.6.SP.4

40. C: To create a box plot, you need five values: the minimum (smallest value), the lower quartile (Q1), the median (Q2), the upper quartile (Q3), and the maximum (largest value). Before you can find these values, first put the data set in order from least to greatest:

12, 16, 17, 17, 18, 18, 19, 20, 20, 20, 21

From this, you can immediately tell that the minimum is 12 and the maximum is 21.

Next find the median (Q2) of the data. Since there are eleven values, the median is the sixth value in the set, 18. This value splits the data set into two halves (not including the median):

12, 16, 17, 17, 18 and 19, 20, 20, 20, 21

The lower quartile (Q1) is the median of the lower half, which is 17, and the upper quartile (Q3) is the median of the upper half, which is 20.

Finally, plot these five values with a box plot using a dot for the minimum (12) and maximum (21) and a box that extends from Q1 (17) to Q3 (18) with a line through it at Q2 (20):

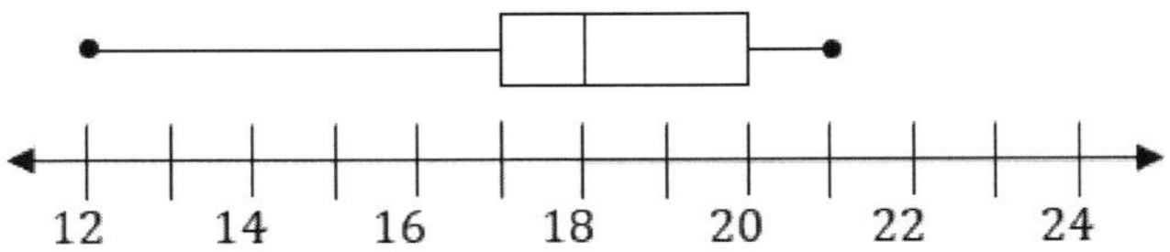

Math.Content.6.SP.5a

41. C: A histogram is a way of graphically representing the distribution of a data set. In a histogram, the data range is divided into intervals so that every value falls in one interval. For the given times, one natural way to set up the intervals is 23 – 23.9, 24 – 24.9, 25 – 25.9, 26 – 26.9, and 27 – 27.9.

Next find the number of values that fall within each interval, i.e. ask how many values are between 23 and 23.9 (including 23 and 23.9), how many are between 24 and 24.9, and so on.

Interval	Frequency
23 – 23.9	2
24 – 24.9	4
25 – 25.9	1
26 – 26.9	2
27 – 27.9	1

Finally, display this information graphically using a vertical bar for each interval and making its height equal to the frequency of the interval:

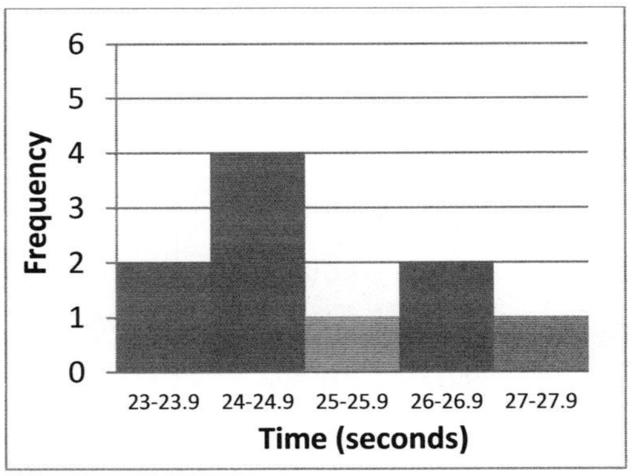

Math.Content.6.SP.5b

42. B: The appropriate unit of measure of a quantity corresponds to what you are counting when you measure it. For instance, if the price of a car is $40,000, because you are counting dollars, the unit of measure of the quantity $40,000 is dollars. To measure the salary of a CEO, you count how much money he or she makes per year. Therefore, the unit of measure is dollars.

Math.Content.6.SP.5c

43. C: The interquartile range of a data set is the difference between the upper and lower quartiles (Q3 and Q1). To calculate it, first put the data set in order from least to greatest:

0, 2, 4, 5, 9, 12, 12

Since there are seven values, the median is the fourth value in the set, 5. This value splits the data set into two halves (not including the median):

0, 2, 4 and 9, 12, 12

The lower quartile is the median of the lower half, which is 2, and the upper quartile is the median of the upper half, which is 12. Subtract these values to find the interquartile range:

$$12 - 2 = 10$$

Math.Content.6.SP.5d

44. B: Data is skewed if values on one side tend to be farther from the 'center' than values on the other side. The graphs below show data skewed to the left and right, respectively.

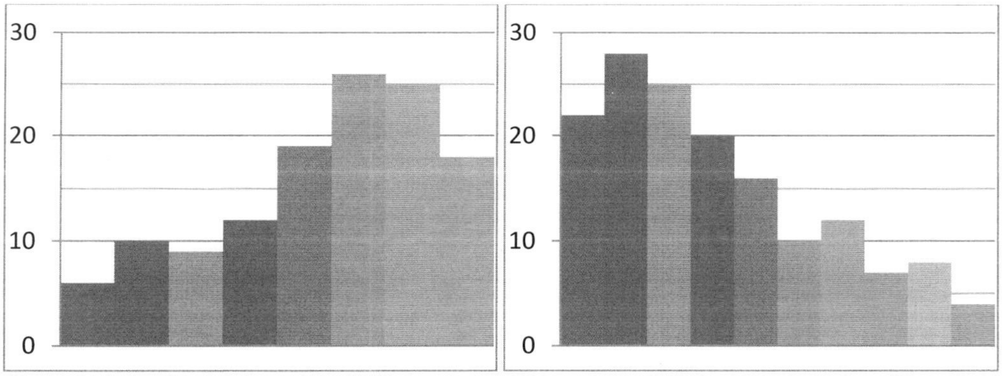

In these situations, the mean get 'pulled away' from what we think of as the more natural value of the center of the data. In the graphs above, the mean gets pulled to the left in the left graph and pulled to the right in the right one. Consequently, in skewed data, the median is usually thought of as a better representative of the center of the data than the mean.

Mode, while a measure of the 'center', is most helpful when the data values are not numeric. Since they are in this case (test scores), the median is still the most useful. (Range is not a measure of center, but a measure of spread.)

How to Overcome Test Anxiety

Just the thought of taking a test is enough to make most people a little nervous. A test is an important event that can have a long-term impact on your future, so it's important to take it seriously and it's natural to feel anxious about performing well. But just because anxiety is normal, that doesn't mean that it's helpful in test taking, or that you should simply accept it as part of your life. Anxiety can have a variety of effects. These effects can be mild, like making you feel slightly nervous, or severe, like blocking your ability to focus or remember even a simple detail.

If you experience test anxiety—whether severe or mild—it's important to know how to beat it. To discover this, first you need to understand what causes test anxiety.

Causes of Test Anxiety

While we often think of anxiety as an uncontrollable emotional state, it can actually be caused by simple, practical things. One of the most common causes of test anxiety is that a person does not feel adequately prepared for their test. This feeling can be the result of many different issues such as poor study habits or lack of organization, but the most common culprit is time management. Starting to study too late, failing to organize your study time to cover all of the material, or being distracted while you study will mean that you're not well prepared for the test. This may lead to cramming the night before, which will cause you to be physically and mentally exhausted for the test. Poor time management also contributes to feelings of stress, fear, and hopelessness as you realize you are not well prepared but don't know what to do about it.

Other times, test anxiety is not related to your preparation for the test but comes from unresolved fear. This may be a past failure on a test, or poor performance on tests in general. It may come from comparing yourself to others who seem to be performing better or from the stress of living up to expectations. Anxiety may be driven by fears of the future—how failure on this test would affect your educational and career goals. These fears are often completely irrational, but they can still negatively impact your test performance.

Elements of Test Anxiety

As mentioned earlier, test anxiety is considered to be an emotional state, but it has physical and mental components as well. Sometimes you may not even realize that you are suffering from test anxiety until you notice the physical symptoms. These can include trembling hands, rapid heartbeat, sweating, nausea, and tense muscles. Extreme anxiety may lead to fainting or vomiting. Obviously, any of these symptoms can have a negative impact on testing. It is important to recognize them as soon as they begin to occur so that you can address the problem before it damages your performance.

The mental components of test anxiety include trouble focusing and inability to remember learned information. During a test, your mind is on high alert, which can help you recall information and stay focused for an extended period of time. However, anxiety interferes with your mind's natural processes, causing you to blank out, even on the questions you know well. The strain of testing during anxiety makes it difficult to stay focused, especially on a test that may take several hours. Extreme anxiety can take a huge mental toll, making it difficult not only to recall test information but even to understand the test questions or pull your thoughts together.

Effects of Test Anxiety

Test anxiety is like a disease—if left untreated, it will get progressively worse. Anxiety leads to poor performance, and this reinforces the feelings of fear and failure, which in turn lead to poor performances on subsequent tests. It can grow from a mild nervousness to a crippling condition. If allowed to progress, test anxiety can have a big impact on your schooling, and consequently on your future.

Test anxiety can spread to other parts of your life. Anxiety on tests can become anxiety in any stressful situation, and blanking on a test can turn into panicking in a job situation. But fortunately, you don't have to let anxiety rule your testing and determine your grades. There are a number of relatively simple steps you can take to move past anxiety and function normally on a test and in the rest of life.

Physical Steps for Beating Test Anxiety

While test anxiety is a serious problem, the good news is that it can be overcome. It doesn't have to control your ability to think and remember information. While it may take time, you can begin taking steps today to beat anxiety.

Just as your first hint that you may be struggling with anxiety comes from the physical symptoms, the first step to treating it is also physical. Rest is crucial for having a clear, strong mind. If you are tired, it is much easier to give in to anxiety. But if you establish good sleep habits, your body and mind will be ready to perform optimally, without the strain of exhaustion. Additionally, sleeping well helps you to retain information better, so you're more likely to recall the answers when you see the test questions.

Getting good sleep means more than going to bed on time. It's important to allow your brain time to relax. Take study breaks from time to time so it doesn't get overworked, and don't study right before bed. Take time to rest your mind before trying to rest your body, or you may find it difficult to fall asleep.

Along with sleep, other aspects of physical health are important in preparing for a test. Good nutrition is vital for good brain function. Sugary foods and drinks may give a burst of energy but this burst is followed by a crash, both physically and emotionally. Instead, fuel your body with protein and vitamin-rich foods.

Also, drink plenty of water. Dehydration can lead to headaches and exhaustion, especially if your brain is already under stress from the rigors of the test. Particularly if your test is a long one, drink water during the breaks. And if possible, take an energy-boosting snack to eat between sections.

Along with sleep and diet, a third important part of physical health is exercise. Maintaining a steady workout schedule is helpful, but even taking 5-minute study breaks to walk can help get your blood pumping faster and clear your head. Exercise also releases endorphins, which contribute to a positive feeling and can help combat test anxiety.

When you nurture your physical health, you are also contributing to your mental health. If your body is healthy, your mind is much more likely to be healthy as well. So take time to rest, nourish your body with healthy food and water, and get moving as much as possible. Taking these physical steps will make you stronger and more able to take the mental steps necessary to overcome test anxiety.

Mental Steps for Beating Test Anxiety

Working on the mental side of test anxiety can be more challenging, but as with the physical side, there are clear steps you can take to overcome it. As mentioned earlier, test anxiety often stems from lack of preparation, so the obvious solution is to prepare for the test. Effective studying may be the most important weapon you have for beating test anxiety, but you can and should employ several other mental tools to combat fear.

First, boost your confidence by reminding yourself of past success—tests or projects that you aced. If you're putting as much effort into preparing for this test as you did for those, there's no reason you should expect to fail here. Work hard to prepare; then trust your preparation.

Second, surround yourself with encouraging people. It can be helpful to find a study group, but be sure that the people you're around will encourage a positive attitude. If you spend time with others who are anxious or cynical, this will only contribute to your own anxiety. Look for others who are motivated to study hard from a desire to succeed, not from a fear of failure.

Third, reward yourself. A test is physically and mentally tiring, even without anxiety, and it can be helpful to have something to look forward to. Plan an activity following the test, regardless of the outcome, such as going to a movie or getting ice cream.

When you are taking the test, if you find yourself beginning to feel anxious, remind yourself that you know the material. Visualize successfully completing the test. Then take a few deep, relaxing breaths and return to it. Work through the questions carefully but with confidence, knowing that you are capable of succeeding.

Developing a healthy mental approach to test taking will also aid in other areas of life. Test anxiety affects more than just the actual test—it can be damaging to your mental health and even contribute to depression. It's important to beat test anxiety before it becomes a problem for more than testing.

Study Strategy

Being prepared for the test is necessary to combat anxiety, but what does being prepared look like? You may study for hours on end and still not feel prepared. What you need is a strategy for test prep. The next few pages outline our recommended steps to help you plan out and conquer the challenge of preparation.

Step 1: Scope Out the Test

Learn everything you can about the format (multiple choice, essay, etc.) and what will be on the test. Gather any study materials, course outlines, or sample exams that may be available. Not only will this help you to prepare, but knowing what to expect can help to alleviate test anxiety.

Step 2: Map Out the Material

Look through the textbook or study guide and make note of how many chapters or sections it has. Then divide these over the time you have. For example, if a book has 15 chapters and you have five days to study, you need to cover three chapters each day. Even better, if you have the time, leave an extra day at the end for overall review after you have gone through the material in depth.

If time is limited, you may need to prioritize the material. Look through it and make note of which sections you think you already have a good grasp on, and which need review. While you are studying, skim quickly through the familiar sections and take more time on the challenging parts.

Write out your plan so you don't get lost as you go. Having a written plan also helps you feel more in control of the study, so anxiety is less likely to arise from feeling overwhelmed at the amount to cover.

Step 3: Gather Your Tools

Decide what study method works best for you. Do you prefer to highlight in the book as you study and then go back over the highlighted portions? Or do you type out notes of the important information? Or is it helpful to make flashcards that you can carry with you? Assemble the pens, index cards, highlighters, post-it notes, and any other materials you may need so you won't be distracted by getting up to find things while you study.

If you're having a hard time retaining the information or organizing your notes, experiment with different methods. For example, try color-coding by subject with colored pens, highlighters, or post-it notes. If you learn better by hearing, try recording yourself reading your notes so you can listen while in the car, working out, or simply sitting at your desk. Ask a friend to quiz you from your flashcards, or try teaching someone the material to solidify it in your mind.

Step 4: Create Your Environment

It's important to avoid distractions while you study. This includes both the obvious distractions like visitors and the subtle distractions like an uncomfortable chair (or a too-comfortable couch that makes you want to fall asleep). Set up the best study environment possible: good lighting and a comfortable work area. If background music helps you focus, you may want to turn it on, but otherwise keep the room quiet. If you are using a computer to take notes, be sure you don't have any other windows open, especially applications like social media, games, or anything else that could distract you. Silence your phone and turn off notifications. Be sure to keep water close by so you stay hydrated while you study (but avoid unhealthy drinks and snacks).

Also, take into account the best time of day to study. Are you freshest first thing in the morning? Try to set aside some time then to work through the material. Is your mind clearer in the afternoon or evening? Schedule your study session then. Another method is to study at the same time of day that you will take the test, so that your brain gets used to working on the material at that time and will be ready to focus at test time.

Step 5: Study!

Once you have done all the study preparation, it's time to settle into the actual studying. Sit down, take a few moments to settle your mind so you can focus, and begin to follow your study plan. Don't give in to distractions or let yourself procrastinate. This is your time to prepare so you'll be ready to fearlessly approach the test. Make the most of the time and stay focused.

Of course, you don't want to burn out. If you study too long you may find that you're not retaining the information very well. Take regular study breaks. For example, taking five minutes out of every hour to walk briskly, breathing deeply and swinging your arms, can help your mind stay fresh.

As you get to the end of each chapter or section, it's a good idea to do a quick review. Remind yourself of what you learned and work on any difficult parts. When you feel that you've mastered the material, move on to the next part. At the end of your study session, briefly skim through your notes again.

But while review is helpful, cramming last minute is NOT. If at all possible, work ahead so that you won't need to fit all your study into the last day. Cramming overloads your brain with more information than it can process and retain, and your tired mind may struggle to recall even previously learned information when it is overwhelmed with last-minute study. Also, the urgent nature of cramming and the stress placed on your brain contribute to anxiety. You'll be more likely to go to the test feeling unprepared and having trouble thinking clearly.

So don't cram, and don't stay up late before the test, even just to review your notes at a leisurely pace. Your brain needs rest more than it needs to go over the information again. In fact, plan to finish your studies by noon or early afternoon the day before the test. Give your brain the rest of the day to relax or focus on other things, and get a good night's sleep. Then you will be fresh for the test and better able to recall what you've studied.

Step 6: Take a Practice Test

Many courses offer sample tests, either online or in the study materials. This is an excellent resource to check whether you have mastered the material, as well as to prepare for the test format and environment.

Check the test format ahead of time: the number of questions, the type (multiple choice, free response, etc.), and the time limit. Then create a plan for working through them. For example, if you have 30 minutes to take a 60-question test, your limit is 30 seconds per question. Spend less time on the questions you know well so that you can take more time on the difficult ones.

If you have time to take several practice tests, take the first one open book, with no time limit. Work through the questions at your own pace and make sure you fully understand them. Gradually work up to taking a test under test conditions: sit at a desk with all study materials put away and set a timer. Pace yourself to make sure you finish the test with time to spare and go back to check your answers if you have time.

After each test, check your answers. On the questions you missed, be sure you understand why you missed them. Did you misread the question (tests can use tricky wording)? Did you forget the information? Or was it something you hadn't learned? Go back and study any shaky areas that the practice tests reveal.

Taking these tests not only helps with your grade, but also aids in combating test anxiety. If you're already used to the test conditions, you're less likely to worry about it, and working through tests until you're scoring well gives you a confidence boost. Go through the practice tests until you feel comfortable, and then you can go into the test knowing that you're ready for it.

Test Tips

On test day, you should be confident, knowing that you've prepared well and are ready to answer the questions. But aside from preparation, there are several test day strategies you can employ to maximize your performance.

First, as stated before, get a good night's sleep the night before the test (and for several nights before that, if possible). Go into the test with a fresh, alert mind rather than staying up late to study.

Try not to change too much about your normal routine on the day of the test. It's important to eat a nutritious breakfast, but if you normally don't eat breakfast at all, consider eating just a protein bar. If you're a coffee drinker, go ahead and have your normal coffee. Just make sure you time it so that the caffeine doesn't wear off right in the middle of your test. Avoid sugary beverages, and drink enough water to stay hydrated but not so much that you need a restroom break 10 minutes into the test. If your test isn't first thing in the morning, consider going for a walk or doing a light workout before the test to get your blood flowing.

Allow yourself enough time to get ready, and leave for the test with plenty of time to spare so you won't have the anxiety of scrambling to arrive in time. Another reason to be early is to select a good seat. It's helpful to sit away from doors and windows, which can be distracting. Find a good seat, get out your supplies, and settle your mind before the test begins.

When the test begins, start by going over the instructions carefully, even if you already know what to expect. Make sure you avoid any careless mistakes by following the directions.

Then begin working through the questions, pacing yourself as you've practiced. If you're not sure on an answer, don't spend too much time on it, and don't let it shake your confidence. Either skip it and come back later, or eliminate as many wrong answers as possible and guess among the remaining ones. Don't dwell on these questions as you continue—put them out of your mind and focus on what lies ahead.

Be sure to read all of the answer choices, even if you're sure the first one is the right answer. Sometimes you'll find a better one if you keep reading. But don't second-guess yourself if you do immediately know the answer. Your gut instinct is usually right. Don't let test anxiety rob you of the information you know.

If you have time at the end of the test (and if the test format allows), go back and review your answers. Be cautious about changing any, since your first instinct tends to be correct, but make sure you didn't misread any of the questions or accidentally mark the wrong answer choice. Look over any you skipped and make an educated guess.

At the end, leave the test feeling confident. You've done your best, so don't waste time worrying about your performance or wishing you could change anything. Instead, celebrate the successful completion of this test. And finally, use this test to learn how to deal with anxiety even better next time.

Review Video: Test Anxiety
Visit mometrix.com/academy and enter code: 100340

Important Qualification

Not all anxiety is created equal. If your test anxiety is causing major issues in your life beyond the classroom or testing center, or if you are experiencing troubling physical symptoms related to your anxiety, it may be a sign of a serious physiological or psychological condition. If this sounds like your situation, we strongly encourage you to seek professional help.

How to Overcome Your Fear of Math

Not again. You're sitting in math class, look down at your test, and immediately start to panic. Your stomach is in knots, your heart is racing, and you break out in a cold sweat. You're staring at the paper, but everything looks like it's written in a foreign language. Even though you studied, you're blanking out on how to begin solving these problems.

Does this sound familiar? If so, then you're not alone! You may be like millions of other people who experience math anxiety. Anxiety about performing well in math is a common experience for students of all ages. In this article, we'll discuss what math anxiety is, common misconceptions about learning math, and tips and strategies for overcoming math anxiety.

What Is Math Anxiety?

Psychologist Mark H. Ashcraft explains math anxiety as a feeling of tension, apprehension, or fear that interferes with math performance. Having math anxiety negatively impacts people's beliefs about themselves and what they can achieve. It hinders achievement within the math classroom and affects the successful application of mathematics in the real world.

Symptoms and Signs of Math Anxiety

To overcome math anxiety, you must recognize its symptoms. Becoming aware of the signs of math anxiety is the first step in addressing and resolving these fears.

Negative Self-Talk

If you have math anxiety, you've most likely said at least one of these statements to yourself:

- "I hate math."
- "I'm not good at math."
- "I'm not a math person."

The way we speak to ourselves and think about ourselves matters. Our thoughts become our words, our words become our actions, and our actions become our habits. Thinking negatively about math creates a self-fulfilling prophecy. In other words, if you take an idea as a fact, then it will come true because your behaviors will align to match it.

Avoidance

Some people who are fearful or anxious about math will tend to avoid it altogether. Avoidance can manifest in the following ways:

- Lack of engagement with math content
- Not completing homework and other assignments
- Not asking for help when needed
- Skipping class
- Avoiding math-related courses and activities

Avoidance is one of the most harmful impacts of math anxiety. If you steer clear of math at all costs, then you can't set yourself up for the success you deserve.

Lack of Motivation

Students with math anxiety may experience a lack of motivation. They may struggle to find the incentive to get engaged with what they view as a frightening subject. These students are often overwhelmed, making it difficult for them to complete or even start math assignments.

Procrastination

Another symptom of math anxiety is procrastination. Students may voluntarily delay or postpone their classwork and assignments, even if they know there will be a negative consequence for doing so. Additionally, they may choose to wait until the last minute to start projects and homework, even when they know they need more time to put forth their best effort.

Physiological Reactions

Many people with a fear of math experience physiological side effects. These may include an increase in heart rate, sweatiness, shakiness, nausea, and irregular breathing. These symptoms make it difficult to focus on the math content, causing the student even more stress and fear.

Strong Emotional Responses

Math anxiety also affects people on an emotional level. Responding to math content with strong emotions such as panic, anger, or despair can be a sign of math anxiety.

Low Test Scores and Performance

Low achievement can be both a symptom and a cause of math anxiety. When someone does not take the steps needed to perform well on tests and assessments, they are less likely to pass. The more they perform poorly, the more they accept this poor performance as a fact that can't be changed.

Feeling Alone

People who experience math anxiety feel like they are the only ones struggling, even if the math they are working on is challenging to many people. Feeling isolated in what they perceive as failure can trigger tension or nervousness.

Feeling of Permanency

Math anxiety can feel very permanent. You may assume that you are naturally bad at math and always will be. Viewing math as a natural ability rather than a skill that can be learned causes people to believe that nothing will help them improve. They take their current math abilities as fact and assume that they can't be changed. As a result, they give up, stop trying to improve, and avoid engaging with math altogether.

Lack of Confidence

People with low self-confidence in math tend to feel awkward and incompetent when asked to solve a math problem. They don't feel comfortable taking chances or risks when problem-solving because they second-guess themselves and assume they are incorrect. They don't trust in their ability to learn the content and solve problems correctly.

Panic

A general sense of unexplained panic is also a sign of math anxiety. You may feel a sudden sense of fear that triggers physical reactions, even when there is no apparent reason for such a response.

Causes Of Math Anxiety

Math anxiety can start at a young age and may have one or more underlying causes. Common causes of math anxiety include the following:

The Attitude of Parents or Guardians

Parents often put pressure on their children to perform well in school. Although their intentions are usually good, this pressure can lead to anxiety, especially if the student is struggling with a subject or class.

Perhaps your parents or others in your life hold negative predispositions about math based on their own experiences. For instance, if your mother once claimed she was not good at math, then you might have incorrectly interpreted this as a predisposed trait that was passed down to you.

Teacher Influence

Students often pick up on their teachers' attitudes about the content being taught. If a teacher is happy and excited about math, students are more likely to mirror these emotions. However, if a teacher lacks enthusiasm or genuine interest, then students are more inclined to disengage.

Teachers have a responsibility to cultivate a welcoming classroom culture that is accepting of mistakes. When teachers blame students for not understanding a concept, they create a hostile classroom environment where mistakes are not tolerated. This tension increases student stress and anxiety, creating conditions that are not conducive to inquiry and learning. Instead, when teachers normalize mistakes as a natural part of the problem-solving process, they give their students the freedom to explore and grapple with the math content. In such an environment, students feel comfortable taking chances because they are not afraid of being wrong.

Students need teachers that can help when they're having problems understanding difficult concepts. In doing so, educators may need to change how they teach the content. Since different people have unique learning styles, it's the job of the teacher to adapt to the needs of each student. Additionally, teachers should encourage students to explore alternate problem-solving strategies, even if it's not the preferred method of the educator.

Fear of Being Wrong

Embarrassing situations can be traumatic, especially for young children and adolescents. These experiences can stay with people through their adult lives. Those with math anxiety may experience a fear of being wrong, especially in front of a group of peers. This fear can be paralyzing, interfering with the student's concentration and ability to focus on the problem at hand.

Timed Assessments

Timed assessments can help improve math fluency, but they often create unnecessary pressure for students to complete an unrealistic number of problems within a specified timeframe. Many studies have shown that timed assessments often result in increased levels of anxiety, reducing a student's overall competence and ability to problem-solve.

Debunking Math Myths

There are lots of myths about math that are related to the causes and development of math-related anxiety. Although these myths have been proven to be false, many people take them as fact. Let's go over a few of the most common myths about learning math.

Myth: Men Are Better at Math Than Women

Math has a reputation for being a male-dominant subject, but this doesn't mean that men are inherently better at math than women. Many famous mathematical discoveries have been made by women. Katherine Johnson, Dame Mary Lucy Cartwright, and Marjorie Lee Brown are just a few of the many famous women mathematicians. Expecting to be good or bad at math because of your gender sets you up for stress and confusion. Math is a skill that can be learned, just like cooking or riding a bike.

Myth: There Is Only One Good Way to Solve Math Problems

There are many ways to get the correct answer when it comes to math. No two people have the same brain, so everyone takes a slightly different approach to problem-solving. Moreover, there isn't one way of problem-solving that's superior to another. Your way of working through a problem might differ from someone else's, and that is okay. Math can be a highly individualized process, so the best method for you should be the one that makes you feel the most comfortable and makes the most sense to you.

Myth: Math Requires a Good Memory

For many years, mathematics was taught through memorization. However, learning in such a way hinders the development of critical thinking and conceptual understanding. These skill sets are much more valuable than basic memorization. For instance, you might be great at memorizing mathematical formulas, but if you don't understand what they mean, then you can't apply them to different scenarios in the real world. When a student is working from memory, they are limited in the strategies available to them to problem-solve. In other words, they assume there is only one correct way to do the math, which is the method they memorized. Having a variety of problem-solving options can help students figure out which method works best for them. Additionally, it provides students with a better understanding of how and why certain mathematical strategies work. While memorization can be helpful in some instances, it is not an absolute requirement for mathematicians.

Myth: Math Is Not Creative

Math requires imagination and intuition. Contrary to popular belief, it is a highly creative field. Mathematical creativity can help in developing new ways to think about and solve problems. Many people incorrectly assume that all things are either creative or analytical. However, this black-and-white view is limiting because the field of mathematics involves both creativity and logic.

Myth: Math Isn't Supposed To Be Fun

Whoever told you that math isn't supposed to be fun is a liar. There are tons of math-based activities and games that foster friendly competition and engagement. Math is often best learned through play, and lots of mobile apps and computer games exemplify this.

Additionally, math can be an exceptionally collaborative and social experience. Studying or working through problems with a friend often makes the process a lot more fun. The excitement and satisfaction of solving a difficult problem with others is quite rewarding. Math can be fun if you look for ways to make it more collaborative and enjoyable.

Myth: Not Everyone Is Capable of Learning Math

There's no such thing as a "math person." Although many people think that you're either good at math or you're not, this is simply not true. Everyone is capable of learning and applying mathematics. However, not everyone learns the same way. Since each person has a different learning style, the trick is to find the strategies and learning tools that work best for you. Some people learn best through hands-on experiences, and others find success through the use of visual aids. Others are auditory learners and learn best by hearing and listening. When people are overwhelmed or feel that math is too hard, it's often because they haven't found the learning strategy that works best for them.

Myth: Good Mathematicians Work Quickly and Never Make Mistakes

There is no prize for finishing first in math. It's not a race, and speed isn't a measure of your ability. Good mathematicians take their time to ensure their work is accurate. As you gain more experience and practice, you will naturally become faster and more confident.

Additionally, everyone makes mistakes, including good mathematicians. Mistakes are a normal part of the problem-solving process, and they're not a bad thing. The important thing is that we take the time to learn from our mistakes, understand where our misconceptions are, and move forward.

Myth: You Don't Need Math in the Real World

Our day-to-day lives are so infused with mathematical concepts that we often don't even realize when we're using math in the real world. In fact, most people tend to underestimate how much we do math in our everyday lives. It's involved in an enormous variety of daily activities such as shopping, baking, finances, and gardening, as well as in many careers, including architecture, nursing, design, and sales.

Tips and Strategies for Overcoming Math Anxiety

If your anxiety is getting in the way of your level of mathematical engagement, then there are lots of steps you can take. Check out the strategies below to start building confidence in math today.

Focus on Understanding, Not Memorization

Don't drive yourself crazy trying to memorize every single formula or mathematical process. Instead, shift your attention to understanding concepts. Those who prioritize memorization over conceptual understanding tend to have lower achievement levels in math. Students who memorize may be able to complete some math, but they don't understand the process well enough to apply it to different situations. Memorization comes with time and practice, but it won't help alleviate math anxiety. On the other hand, conceptual understanding will give you the building blocks of knowledge you need to build up your confidence.

Replace Negative Self-Talk with Positive Self-Talk

Start to notice how you think about yourself. Whenever you catch yourself thinking something negative, try replacing that thought with a positive affirmation. Instead of continuing the negative thought, pause to reframe the situation. For ideas on how to get started, take a look at the table below:

Instead of thinking...	Try thinking...
"I can't do this math." "I'm not a math person."	"I'm up for the challenge, and I'm training my brain in math."
"This problem is too hard."	"This problem is hard, so this might take some time and effort. I know I can do this."
"I give up."	"What strategies can help me solve this problem?"
"I made a mistake, so I'm not good at this."	"Everyone makes mistakes. Mistakes help me to grow and understand."
"I'll never be smart enough."	"I can figure this out, and I am smart enough."

Practice Mindfulness

Practicing mindfulness and focusing on your breathing can help alleviate some of the physical symptoms of math anxiety. By taking deep breaths, you can remind your nervous system that you are not in immediate danger. Doing so will reduce your heart rate and help with any irregular breathing or shakiness. Taking the edge off of the physiological effects of anxiety will clear your mind, allowing your brain to focus its energy on problem-solving.

Do Some Math Every Day

Think about learning math as if you were learning a foreign language. If you don't use it, you lose it. If you don't practice your math skills regularly, you'll have a harder time achieving comprehension and fluency. Set some amount of time aside each day, even if it's just for a few minutes, to practice. It might take some discipline to build a habit around this, but doing so will help increase your mathematical self-assurance.

Use All of Your Resources

Everyone has a different learning style, and there are plenty of resources out there to support all learners. When you get stuck on a math problem, think about the tools you have access to, and use them when applicable. Such resources may include flashcards, graphic organizers, study guides, interactive notebooks, and peer study groups. All of these are great tools to accommodate your individual learning style. Finding the tools and resources that work for your learning style will give you the confidence you need to succeed.

Realize That You Aren't Alone

Remind yourself that lots of other people struggle with math anxiety, including teachers, nurses, and even successful mathematicians. You aren't the only one who panics when faced with a new or challenging problem. It's probably much more common than you think. Realizing that you aren't alone in your experience can help put some distance between yourself and the emotions you feel about math. It also helps to normalize the anxiety and shift your perspective.

Ask Questions

If there's a concept you don't understand and you've tried everything you can, then it's okay to ask for help! You can always ask your teacher or professor for help. If you're not learning math in a traditional classroom, you may want to join a study group, work with a tutor, or talk to your friends. More often than not, you aren't the only one of your peers who needs clarity on a mathematical concept. Seeking understanding is a great way to increase self-confidence in math.

Remember That There's More Than One Way To Solve A Problem

Since everyone learns differently, it's best to focus on understanding a math problem with an approach that makes sense to you. If the way it's being taught is confusing to you, don't give up. Instead, work to understand the problem using a different technique. There's almost always more than one problem-solving method when it comes to math. Don't get stressed if one of them doesn't make sense to you. Instead, shift your focus to what does make sense. Chances are high that you know more than you think you do.

Visualization

Visualization is the process of creating images in your mind's eye. Picture yourself as a successful, confident mathematician. Think about how you would feel and how you would behave. What would your work area look like? How would you organize your belongings? The more you focus on something, the more likely you are to achieve it. Visualizing teaches your brain that you can achieve whatever it is that you want. Thinking about success in mathematics will lead to acting like a successful mathematician. This, in turn, leads to actual success.

Focus on the Easiest Problems First

To increase your confidence when working on a math test or assignment, try solving the easiest problems first. Doing so will remind you that you are successful in math and that you do have what it takes. This process will increase your belief in yourself, giving you the confidence you need to tackle more complex problems.

Find a Support Group

A study buddy, tutor, or peer group can go a long way in decreasing math-related anxiety. Such support systems offer lots of benefits, including a safe place to ask questions, additional practice with mathematical concepts, and an understanding of other problem-solving explanations that may work better for you. Equipping yourself with a support group is one of the fastest ways to eliminate math anxiety.

Reward Yourself for Working Hard

Recognize the amount of effort you're putting in to overcome your math anxiety. It's not an easy task, so you deserve acknowledgement. Surround yourself with people who will provide you with the positive reinforcement you deserve.

Remember, You Can Do This!

Conquering a fear of math can be challenging, but there are lots of strategies that can help you out. Your own beliefs about your mathematical capabilities can limit your potential. Working toward a growth mindset can have a tremendous impact on decreasing math-related anxiety and building confidence. By knowing the symptoms of math anxiety and recognizing common misconceptions about learning math, you can develop a plan to address your fear of math. Utilizing the strategies discussed can help you overcome this anxiety and build the confidence you need to succeed.

Thank You

We at Mometrix would like to extend our heartfelt thanks to you, our friend and patron, for allowing us to play a part in your journey. It is a privilege to serve people from all walks of life who are unified in their commitment to building the best future they can for themselves.

The preparation you devote to these important testing milestones may be the most valuable educational opportunity you have for making a real difference in your life. We encourage you to put your heart into it—that feeling of succeeding, overcoming, and yes, conquering will be well worth the hours you've invested.

We want to hear your story, your struggles and your successes, and if you see any opportunities for us to improve our materials so we can help others even more effectively in the future, please share that with us as well. **The team at Mometrix would be absolutely thrilled to hear from you!** So please, send us an email (support@mometrix.com) and let's stay in touch.

If you'd like some additional help, check out these other resources we offer for your exam: http://MometrixFlashcards.com/CCSS

Online Resources

Due to our efforts to try to keep this book to a manageable length, we've created a link that will give you access to all of your online resources:

mometrix.com/resources719/ccg6math

It's Your Moment, Let's Celebrate It!

Share your story @mometrixtestpreparation